PRAISE FOR *WHEN WAS JESUS REALLY BORN?*

Weaving scientific and historical scholarship with theological and pastoral concern, Dr. Ware's book provides a fresh, rigorous, and engaging new approach to an old issue. Although we must remain uncertain about the answer to the question the book's title asks, *When Was Jesus Really Born?* offers new clarity about how Christians have seen time and history as revealing God's presence and activity. This book deserves to be widely read by students and scholars, and by pastors and others exploring their faith.

—Rev. Professor Andrew McGowan
Warden, Trinity College, The University of Melbourne
Joan Munro Professor of Historical Theology, Trinity College Theological School

History really matters for Christians. Our faith is founded upon events from the past that really happened. The dates of these historical events have been so disputed that they have created many contradictory calendars, including the Western Gregorian calendar we use today.

Steven Ware focuses upon the dates of key events of the life of Jesus. The birth of Jesus of Nazareth—unlike the legendary births of many rulers of the ancient world—is essential to date, as an affirmation of the full reality of Jesus as a historical, rather than a merely mythological or spiritual, character. If Jesus was a real human being, then he must have been born on a specific date.

Upon what *historical evidence* does this claim of the true humanity of the person Jesus of Nazareth rely? This is the place where Ware's painstaking labors come to our aid. *When Was Jesus Really Born?* offers a truly comprehensive reading of all of the calendars, timelines, church chronologies, and other materials that provide the evidence for dating the birth of our Lord. Ware shows how the calendar dates of Jesus' conception, birth, death, and resurrection are inextricably connected with each other. With special emphasis upon the history of Christianity, Ware also creatively portrays the history of timekeeping from ancient history to the present. This is a daunting and remarkable task.

Would you like to know what the biblical, historical, astronomical, and calendrical evidence shows about the dating of the life of Jesus of Nazareth? Then Ware's book is essential to answer your questions. So, if you really want to know why December 25, 3 BCE may likely be the most viable date for the birth of Jesus, according to Ware and the Western calendar, *read this book.*

—Charles J. Scalise
Professor of Church History
Fuller Theological Seminary

"The most important event in human history," is what Steven Ware the *Christian* called the birth of Jesus Christ. With his particular thoroughness and diligence, Steven Ware the *scientist* considered the question of when it all took place. Broad in scope, he pursues the topic purposefully.

For me, the name and personality of an author are synonyms of profound education, scientific thoroughness, and Christian conviction. Dr. Ware brings together his knowledge of the Bible, history, astronomy, the calendar, and chronometry to show us when Jesus was really born. The comprehensive, informative appendix should not go unmentioned.

I hope that this book will attract the deserved attention.

—Prof. Dr. Manfred Schukowski
Rostock, Germany
Retired Professor of History of Science
Author of *Wunderuhren: Astronomische Uhren in Kirchen der Hansezeit* (Schwerin: Thomas Helms, 2006), *Sonne, Mond, und zwölf Apostel: Die Astronomische Uhr in der Marienkirche zu Rostock* (Schwerin: Thomas Helms, 2012)

Debates on the precise dates of Jesus' birth, his Passion, and Resurrection have been taking place since the early days of Christianity. Though we may not realize it, the Christian calendar affects the schedules and daily lives of most people throughout the world. The puzzles involved in synchronizing cosmic events within a rational calendar have stimulated the development of historical chronology in many ways. Steven Ware's very readable book explains the often complex issues and sets them in a wide historical horizon. His arguments for the dates in question—brought forward by a dedicated Christian—are persuasive, though they will certainly not bring the debates to an end.

—Prof. Dr. Gerhard Dohrn-van Rossum
Professor of Medieval History
Technical University of Chemnitz, Germany
Author of *History of the Hour: Clocks and Modern Temporal Orders*
(Chicago: University of Chicago Press, 1998)

When Was Jesus Really Born?

When Was Jesus Really Born?

Early Christianity, the Calendar, and the Life of Jesus

Steven L. Ware

Peer Reviewed

CONCORDIA PUBLISHING HOUSE · SAINT LOUIS

Peer Reviewed

Published 2013 by Concordia Publishing House
3558 S. Jefferson Ave., St. Louis, MO 63118-3968
1-800-325-3040 · www.cph.org

Manufactured in the United States of America

Library of Congress Cataloging-in-Publication Data

Ware, Steven L.

When was Jesus really born? : early Christianity, the calendar, and the life of Jesus / Steven L Ware.

pages cm

Includes bibliographical references and index.

ISBN 978-0-7586-4197-7

1. Jesus Christ--Date of birth. 2. Jesus Christ--Biography. 3. Church history--Primitive and early church, ca. 30-600. 4. Calendar, Gregorian--History. 5. Calendars. 6. Time measurements. I. Title.

BT318.W37 2013

232.9'01--dc23 2013009522

---

1 2 3 4 5 6 7 8 9 10 21 20 19 18 17 16 15 14 13

CONTENTS

List of Illustrations

PREFACE

"All true research and writing is autobiography," my dissertation mentor once told me. The further my dissertation progressed, the more I discovered the truth of his statement. And although the present topic is far afield from that of my dissertation of years ago, my old friend might be pleased to know that his statement is now even more true than it was previously.

An item of my personal history which my parents shared with me at an early age was the fact that I had been born early on Easter Sunday morning. With a birthday in early April, I therefore assumed that it would coincide with Easter Sunday quite regularly. When I later discovered the error in my assumption and queried my parents about it, they simply answered that Easter had a habit of hopping all around the calendars of March and April. Then one Sunday morning in the 1960s when my restless adolescent mind was having perhaps as much success in listening to the sermon as Mister Bean in his later comedy sketch, I happened to be paging through our worship book, *The Lutheran Hymnal*, when I made the unexpected discovery which began my long journey toward the current project.

Near the end of all the liturgical details at the beginning of the text (only 170 pages) and just before the hymns, under the heading "Miscellaneous," my eyes found delight and my mind began to find intellectual sustenance in lists and charts which were anything but miscellaneous matters to me. The first was the "Table of the Days on which Easter will Fall from 1941–2000," followed by "A Table of the Movable Feasts and Festivals" and "A Table of Lessons for the Sundays, Feasts, and Chief Festivals of the Church Year."[1] With no formal knowledge of Latin at the time, I was intrigued by the many

[1] *The Lutheran Hymnal* (St. Louis: Concordia, 1941), 158ff.

Latin titles for Sundays in the church year, perhaps most notably by "Sexagesima Sunday."[2]

Aside from an introduction to the details of the ecclesiastical calendar, however, other bits of more personally relevant information became immediately apparent from even a cursory glance at the first two "tables." The first was that Easter had occurred on my birth date eleven years before my birth—a pattern likewise followed with some other dates in late March and April. The second and less encouraging observation was that Easter would not again fall on the calendar date of my birth until sometime after the beginning of the twenty-first century. In addition, I learned from the second table concerning "the several days that Easter can possibly fall upon" that "several" had been stretched well beyond its usual limits to encompass the thirty-five possible dates for (western) Easter between 22 March and 25 April.

This book is about much more than the proper date(s) of Easter, of course. As the title suggests, it begins with an exploration of the date of Jesus' birth. But the reader will soon become aware that the calendrical dates of Jesus' birth, death, and resurrection are inextricably linked, and form one of the fascinating stories of early Christianity. As subsequent pages will show, I am far from the first to investigate the question of the dates of Jesus' birth, death, and resurrection. In fact, this question and the related matters of time measurement have consumed some of the greatest minds of human history, and it is therefore with a palpable sense of humility that I have dared to enter this discussion.

In addition, I must gladly concede that I have not engaged in this investigation alone, but that my queries have been refined and augmented along the way by numerous other individuals who deserve recognition for their contributions. First of all, heartfelt thanks are due to the President and administration of Nyack College and Alliance Theological Seminary, whose generous support of this project in the form of both a sabbatical and a summer writing grant were essential

[2] Sexagesima Sunday is the second to last Sunday in the ecclesiastical calendar before Ash Wednesday. "Sexagesima" means "sixtieth," referring to the approximately sixty days between this Sunday and Easter Sunday (actually 57). Quinquagesima ("fiftieth"), the last Sunday before Ash Wednesday (also known as the Feast of the Transfiguration), is mathematically correct as standing fifty days from Easter. See http://www.newadvent.org/cathen/13747a.htm.

to my ability to travel on multiple occasions to multiple countries and continents, and consequently to complete the necessary research. Secondly, I should recognize and thank the many students who have attended my courses at both campuses of Nyack College in New York City and Nyack, NY, for their patience in observing my progressive development of the material of this book over the past several years, and for their many questions and comments which pushed me to improve and clarify its message. Along with them, I must thank several of my colleagues among the Nyack faculty who have questioned and debated pieces of this material with me. Foremost among them is David A. Weir, Ph.D., Professor of History, whose interest in my topic, suggestions of additional resources, and relentless critique of numerous fine points in the manuscript added incalculable value to its final form.

Thanks are likewise due to the interlibrary loan staff at the Bailey and Wilson libraries of Nyack College for their patient efforts to locate numerous obscure and hard-to-find texts and journal articles from centuries past. Thanks are equally due to J. James Mancuso, Librarian at Northeastern Baptist College, for his many suggestions of additional materials and research assistance in finding them. Jim's enthusiasm for this topic was rather naturally spurred by the fact that he also has a birthday within the range of possible dates for Easter, and furthermore because Easter occurs on the calendrical date of his birth more often than any other.

Thanks are due to individuals at a variety of churches, museums, and state and national parks—in several countries and across several continents—for their assistance in transportation, their enthusiastic sharing of information, and their permission to view and photograph precious artifacts and monuments. In particular, thanks are due to the special events officers at the Vatican Museum for granting special permission and a personal escort into the closed Pio Christian gallery for the purpose of viewing and photographing the sarcophagus of Hippolytus, without which my research would have been sadly incomplete. Thanks are also due to Uta Jahnke and Dr. Manfred Schukowski for their very thorough explanation of the operations of the astronomical clock at Marienkirche (Saint Mary's Church) in Rostock, Germany. Their explanations not only confirmed some of the conclusions made in this text about the historical evolution of the

Easter (Paschal) cycle, but opened doors for further research and future academic collaboration.

My heartfelt thanks are equally due to the editors and production assistants at Concordia Publishing House, without whose partnership this project would not have come to its final form. Thanks are particularly due to the Reverend Edward A. Engelbrecht, (former) Senior Editor for Professional and Academic Books, whose initial interest in my manuscript launched its long journey through the publication process. Thanks are also due to Laura Lane, Associate Editor of Professional and Academic Books, who guided the proposal through the necessary approvals. And thanks are equally due to Sarah Steiner, Production Editor for Professional and Academic Books, whose insights and very patient answers to my many questions were always accurate, helpful, and knowledgeable. In addition, thanks are due to the Rev. Dr. Larry Myers who, along with the aforementioned Dr. David Weir, served as a peer reviewer of the manuscript and thereby added immeasurably to its value.

Finally, it must be admitted that mere words are woefully inadequate to express my deepest heartfelt thanks to my wife Rebecca, to whom this book is most deservingly dedicated. She has been an absolutely faithful companion in all of life's endeavors for more than thirty years, and a wonderful mother to our three children. Consequently, she has struggled with me through the long evolution and maturation of the topics contained in this book. She has been a most enthusiastic travel and research partner to literally hundreds of locations in dozens of countries on five continents. (It was she who refused to let me give up when we discovered that the Pio Christian collection at the Vatican Museum was closed!) And she has never been afraid to bring my far-ranging ruminations back to earth with the simple question of their relevance. I can never thank her enough.

My debt to my wife is exceeded only by my debt to Almighty God, who loved me enough not only to pay the debt of my sin through the death and resurrection of His Son Jesus Christ, but also to choose me as one of His own, give me the strangest of all birthdays, and then challenge me to somehow figure it out.

New York City
11 January 2013

RESOURCE ABBREVIATIONS

Adv. Haer.	Irenaeus, *Against Heresies*
AJP	*The American Journal of Philology*
ANF	*The Ante-Nicene Fathers.* Eds. A. Cleveland Coxe, Alexander Roberts, James Donaldson, Philip Schaff, and Henry Wace. 10 vols. Edinburgh: T & T Clark, 1885–1887.
Ant.	Josephus, *Jewish Antiquities*
BJRL	*Bulletin of the John Rylands University Library of Manchester*
CKC	*Chronos, Kairos, Christos.* Eds. Jerry Vardaman and Edwin M. Yamauchi. Winona Lake, IN: Eisenbrauns, 1989.
EJ	*Encyclopedia Judaica.* Ed. Cecil Roth. 18 vols. Jerusalem: Keter/Macmillan, 1978.
JBL	*Journal of Biblical Literature*
JETS	*Journal of the Evangelical Theological Society*
JTS	*Journal of Theological Studies*
LW	*Luther's Works.* American Edition. 55 vols. Jaroslav Pelikan, ed. St. Louis: Concordia, 1958.
MGH	*Monumenta Germaniae Historica.*
NPNF	*A Select Library of the Nicene and Post-Nicene Fathers of the Christian Church.* Ed. Philip Schaff and Henry Wace. Two series. 28 vols. Edinburgh: T & T Clark, 1886–1900.

OCY	Blackburn, Bonnie, and Leofranc Holford-Strevens. *The Oxford Companion to the Year: An exploration of calendar customs and time-reckoning.* Oxford: University Press, 1999.
ODCC	*The Oxford Dictionary of the Christian Church.* Ed. Frank L. Cross and E. A. Livingstone. Third edition. Oxford: University Press, 2005.
PG	*Patrologia Graeca. Patrologiae cursus completus: series Graeca.* Ed. J. P. Migne, et al. 162 vols. Paris: 1857–1867.
PL	*Patrologia Latina. Patrologiae cursus completus: series Latina.* Ed. J. P. Migne, et. al. 221 vols. Paris: 1844–1902.
TWOT	Waltke, Bruce K., R. Laird Harris, and Gleason L. Archer. *Theological Wordbook of the Old Testament.* 2 vols. Chicago: Moody Press, 1980.

INTRODUCTION

A SEASON FOR REASON

WHY A BOOK ON EARLY CHRISTIANITY, THE CALENDAR, AND THE LIFE OF JESUS?

The modern dilemma concerning Jesus' birth and celebrations thereof was illustrated sharply in the New York metropolitan area in November 2010. At the New Jersey entrance to the Holland Tunnel a billboard was leased by American Atheists with the message "You know it's a myth. This season, celebrate reason." The image behind the message was that of the wise men following the star and riding their camels toward Bethlehem. Spokespersons for American Atheists claimed that they were not making an anti-Christian or anti-Christmas statement, but that they merely wished to "celebrate reason," as the sign stated. One might wonder, however, why they chose not to illustrate their billboard with a dreidel and menorah or other images of the Jewish feast of Chanukah being celebrated at the same season, or why they chose not to illustrate their sign with images of the Muslim feast of Ramadan, celebrated just weeks previously. Fortunately, the tension was somewhat defused when the Catholic League put up a billboard on the Manhattan side of the tunnel stating, "You know it's real. This season, celebrate Jesus."

Indeed, in our modern post-Christian society, matters concerning the birth of the historical person Jesus and Christian claims about the uniqueness of the circumstances of his birth are relegated to the realm of faith and myth—widely respected, yet also widely disrespected and ignored. Even though many churches typically see their largest crowds of worshipers on Christmas Eve (or on Easter Sunday), a large percentage of their neighbors are just as happy if no religious meaning at all is attached to what is otherwise a holiday full of feasting and materialistic pleasures. In fact, many moderns would greatly prefer that Christians would keep their faith to themselves as a purely private matter and not engage in large and highly publicized worship events to mark the holiday, since talk of a Messiah sent to save humanity carries with it the uncomfortable moral message that humans are sinners who actually need a Messiah.

For the vast majority of Christians, however, who prefer not to relegate their faith to the realm of myth or confine their faith to a privatistic box to be opened only in the friendly confines of like-minded individuals, this is a matter to be treated with utmost seriousness and care. To a limited degree, the dismissive and impatient reactions are readily understandable, given the voluminous and sometimes divisive history of discussions regarding the birth of Jesus and the celebration thereof. Yet like all great issues of pivotal significance in human history, it is not possible or even advisable to close the discussion, especially when new and previously unattended information offers to shed new light on an old issue.

In this book the author seeks to engage a foundational question behind Christian faith and practice for most of the past 2,000 years—the date of the birth of the Messiah, the Lord and Savior Jesus Christ. As succeeding chapters will show, however, the equally engaging question for Christians is that of the likely dates of his death and resurrection, and how the celebration of Jesus' resurrection developed into the Paschal, or Easter cycle we know today. In attempting to locate these dates, this author is dependent upon the historical clues given in the Bible, and likewise upon the chronological assertions of authors—especially Christian authors—during the first several centuries of the Christian era. Indeed, it is the assumption of this author that a most fruitful avenue of investigation for this issue is to consult those who were among the first to confront these matters.

Consequently, the conclusions regarding the chronological dates of Jesus' life which are drawn by this investigation will be twofold: First of all, we will endeavor to establish the most likely calendar dates of Jesus' life, based on biblical, historical, and calendrical evidence. Secondly, we will follow the development of early Christian calculations of the essential dates of Jesus' life, especially with regard to their understanding of time, calendars, and the divine redemptive plan. Moreover, we shall see that although we cannot often claim scientific accuracy or historical certitude for early Christian calculations, we can nevertheless assert that they are genuinely "Christian" and ancient answers to the question. And by their status as certifiably ancient Christian answers, it is secondarily hoped that they offer viable insights on this matter of historic import.

The interested reader who takes the time to browse through the literature regarding the birth date of Jesus or executes an online search is immediately confronted with a plethora of general interest articles, as well as other articles representing a wide variety of perspectives and sub-categories of the larger issue. For to do this is to enter into an arena which has engaged many great minds of centuries past and present who have given it serious debate on biblical, historical, and astronomical grounds, and in whose presence this author enters with deep humility and reverence. In fact, when one considers not only the volume of the material, but especially the sources of information used by early Christians and the manner in which they constructed their understanding of the chronological details of the life of Jesus, it becomes apparent that it is Christians who have consistently employed reason in this matter while atheists have simply substituted "reason" for non-belief. Therefore, mindful of the greatness of his company compared with his own minute contribution, it is not the intention of this author to offer final answers to the many matters of debate, although a brief review of some of those matters is given in succeeding chapters. It is hoped, rather, that renewed attention to the questions surrounding the calendar dates of Jesus' life may shed new light on the world of early Christianity.

A Brief Overview

This book is about the birth of Jesus. Yet it is also about his resurrection, early Christianity, and the historical evolution of our calendar. The first chapter provides a summary of the variety of issues and debates surrounding the birth date of the historical person of Jesus. What historical clues are given in the Bible and other sources concerning people, places, or situations at the time of Jesus' birth? How do those clues fit into the calendars of the period?

A short description of this author's thoughts concerning early Christian dating for the birth of Jesus is given, although additional reading in later chapters is necessary for a fuller understanding of the matter. In addition, the reader will be introduced to two matters of historical debate which have engaged both Christians and non-Christians: The first is Saturnalia, an ancient pagan celebration of the birthday of the sun occurring in late December, which some have claimed was simply co-opted by Christians and overlaid with the celebration of Jesus' birth. The second is that of the *World Week*—a belief that the entirety of God's redemptive plan for human history is to be accomplished within the span of six thousand years.

Chapter Two begins the exploration of the larger issues involved when asking the question of the birth date of Jesus—the matter of timekeeping. It starts with the experience of our early ancestors—the first farmers—whose existence was dominated first and foremost by the necessity of procuring or producing enough food for survival to the next day, the next season, and the next year. Why was time measurement such an important matter for those living such an apparently simple existence? How is it that ancient civilizations came up with numerous stunning advancements in time measurement? What were the first calendars like? And how was time measurement linked to religious observances?

Chapter Three will observe some of the basic facts of earth science and astronomy which were slowly grasped by our ancestors in learning to measure time. Why is it that the task of time measurement was so difficult for so long? What is meant by terms such as "ecliptic," "Tropical Year," or "Sidereal Year"? How are these related to the construction of calendars? And finally, are the multiple astronomical factors of imprecision in time measurement perhaps an indication of divine origin of the cosmos?

Chapter Four focuses on the rise of Christianity and its unmistakable impact on the measurement of time. While most moderns are aware of the fact that the western chronological system is based on the approximate date of the birth of Jesus—hence the title of this text—we will see that the more important matter for early Christians was determining the proper date for the celebration of Jesus' death and resurrection from the dead. Furthermore, we will see how Christians disagreed among themselves over the proper date for celebrating Jesus' resurrection. We will again confront the biblical input on the date of Jesus' resurrection, and how the Jewish calendar impacted early Christian celebrations of the resurrection. Moreover, we will observe early Christian conceptions of world history and their linkage to debates over the proper date for the celebration of Jesus' resurrection.

Chapter Five focuses on the person at the center of most discussions of Christianity and chronology—the sixth-century monk Dionysius Exiguus. An often-ignored but truly influential scholar of the early medieval period, it was he who designed the western or Christian chronological system under which most of the world presently dates itself. Understandably, his work received increased attention in recent years with the approach of the twenty-first century and third Christian millennium. He has been critiqued rather heavily in recent years with regard to both his chronological calculations and his theological perspectives. This chapter will seek to answer such questions as exactly what he was attempting to accomplish and why. Was he aware of his apparent error in calculating the birth date of Jesus? Was he even aware of the latest advances in chronological calculation prior to his own work? What was the impact of Dionysius' Paschal cycle and invention of the "Christian Era"? And finally—based on knowledge of the work of Dionysius and early Christian understandings of calendars—we will return to some of the matters of debate in chapter one concerning the essential dates of Jesus' life.

Chapter Six tells the story of the Christian role in more recent advances in timekeeping. Beginning with the often-misinterpreted trial of Galileo Galilei by Roman Catholic authorities in 1633, it investigates the role of Christians, and especially of the Roman Catholic Church, in scientific advances in timekeeping both before and after that tragic event. We will look backward at the calendar revision of 1582 initiated by Pope Gregory XIII and its gradual adoption by nations both Catholic and Protestant. Why was this revision necessary, and what was its connection to the celebration of Jesus' resurrection? We will also look forward past the trial of Galileo to the Church's further patronage of the science of time measurement in the construction of *meridiane*—simple lines for marking the solar position—which were placed in several cathedrals. Why this further patronage, even after the calendrical revision? And what does this say about the popular perception of the relationship between Christianity and scientific endeavors?

The Epilogue considers one of the odd consequences of the Gregorian calendrical revision—the development of All Fools', or April Fools' Day. It also takes us to the future, if only momentarily, to consider lingering questions with regard to Jesus' birth and chronology, with a few brief thoughts on the ongoing popularity of eschatology among Christians: Does the beginning of the current millennium have anything to do with the date of Jesus' return? If not, are there perhaps other dates of Christological significance coming soon which should be celebrated? More importantly, how should Christians view themselves in relation to time?

As the reader can see by this point, this book is not intended as a Bible commentary or devotional guide to the birth narratives of Jesus in the gospels of Matthew and Luke. Nor is it an attempt at a general history of time measurement. That task has been accomplished rather adequately already by several earlier authors.[1] Nor is this book about time as it regards debates about the absolute age of our earth or the universe, although at least a few such calculations will quite naturally be noted as we deal with the issues connected to an investigation of Christmas, Easter, and ancient calendars.[2] Nor is this primarily a text of astronomy and earth science, although one chapter is devoted to a few of the basic facts of that discipline which form one of the foundations of this study.

Nor does this book attempt to delve into scientific questions regarding the nature of time beyond our immediate experience of it on planet Earth—the realm of Newton's mechanics, Einstein's special and general theories of relativity, or quantum mechanics.[3] Nor is this a study of the evolution of time-keeping instruments, although some

---

[1] The true compendium in this regard is Bonnie Blackburn and Leofranc Holford-Strevens, *The Oxford Companion to the Year: An Exploration of Calendar Customs and Time-Reckoning* (Oxford: University Press, 1999). Other valuable treatments are David Ewing Duncan, *The Calendar* (London: Fourth Estate, 1999), E. G. Richards, *Mapping Time: The Calendar and Its History* (Oxford: University Press, 1998), and Duncan Steel, *Marking Time: The Epic Quest to Invent the Perfect Calendar* (New York: John Wiley, 2000).

[2] A worthwhile text for Christians in this regard is *Three Views on Creation and Evolution* (J. P. Moreland and John Mark Reynolds, eds. Grand Rapids: Zondervan, 1999). As the title indicates, it compares and contrasts the variety of views Christians have taken on this matter.

[3] Those with interests in this field should read the worthwhile treatments of Stephen Hawking, *A Brief History of Time* (New York: Bantam Books, 1988); Peter Coveney and Roger Highfield, *The Arrow of Time: A Voyage Through Science to Solve Time's Greatest Mystery* (New York: Fawcett Columbine, 1990); and Peter Galison, *Einstein's Clocks, Poincaré's Maps: Empires of Time* (New York: W. W. Norton, 2003).

of the individuals and groups mentioned herein were centrally involved in their creation.[4] Nor will this text dedicate large amounts of space to the wide variety of non-Christian calendars in the ancient world or at the present, although occasional mention of their contributions to the progress of accurate time measurement, and especially their contributions to the Christian story, is quite naturally included.[5] As the reader will undoubtedly realize before turning very many pages, to a small extent it could be said that this book is about a particular understanding of human history—the Christian understanding of history—although in the end it is really more about Christians and their attempts to understand time itself.[6]

Numerous times this author has told himself that he should have written this book twenty years ago. At that time, as humanity was approaching the end of the twentieth century and of the second Christian millennium, the attention of many people—although for most only momentarily—was naturally focused on the historical question of the precise end of that century and millennium and the beginning of the next. At the end of the twentieth century some were also asking the religious question of whether there was any cosmic significance to the big change in numbers—whether the new millennium was a harbinger of the imminent return of Jesus (or of some other Messiah), of earth-shaking cataclysms, or of a new world political order. Perhaps my publisher would have appreciated the boost in sales engendered by such fortuitous timing. Aside from the answers for the long delay which are matters of personal and professional history, however, it is hoped that a long engagement with this topic will yield fruitful results which would not have been part of the discussion in the late twentieth century, at least not by this author.

---

[4] See Gerhard Dohrn-van Rossum, *History of the Hour: Clocks and Modern Temporal Orders* (Chicago: University of Chicago Press, 1996); Dava Sobel, *Longitude: The Story of a Lone Genius Who Solved the Greatest Scientific Problem of His Time* (New York: Penguin, 1996).

[5] See Anthony Aveni, *Empires of Time: Calendars, Clocks, and Cultures* (New York: Basic Books, 1989); Manfred Schukowski, *Wunderuhren: Astronomische Uhren in Kirchen der Hansezeit.* Schwerin: Thomas Helms Verlag, 2006); Richards, *Mapping Time*, Part II; Masayuki Sato, "Comparative Ideas of Chronology," *History and Theory* 30, No. 3 (October 1991): 275–301.

[6] Helpful guides in this regard include Robert G. Clouse, Robert N. Hosack, and Richard V. Pierard, *The New Millennium Manual: A Once and Future Guide* (Grand Rapids: Baker, 1999), and *The Meaning of the Millennium: Four Views* (Robert G. Clouse, ed. Downers Grove, IL: InterVarsity, 1977).

Furthermore, it is hoped that the value of this topic, especially for Christians, will be seen to extend far beyond immediate chronological and eschatological questions.

A glance at the bibliography at the conclusion of this text will reveal that this is far from the first attempt to investigate the essential dates of the birth of Jesus, or the conjunction of chronology and Christianity, despite the apparently arcane nature of the subject. The advent of the internet has given a whole new venue for discussion of these matters, as it has for everything else as well. The brilliance of numerous earlier authors in matters of chronology, astronomy, mathematics, liturgical observance, and political and theological history provide benchmarks to which the present writer can only aspire. Many of them are referenced in the following chapters where they provided invaluable assistance. Yet what seems to have been muted or even discounted in some recent discussions of chronology, or even of Christian celebrations such as Christmas and Easter, is the profoundly Christian story of true believers—even covered as they were with the warts of human self-interest and limitations—who were recipients and vessels of divine grace making an impact on their world. It is that story of faith, alongside the matters of academic investigation, to which the reader is invited herein.

One lesson learned quickly by most of us who have engaged in the world of academic research and writing is that there is rarely a *final* word on anything. Regardless of how exhaustive and comprehensive our research, and in spite of the apparent persuasiveness of our arguments and conclusions, it is inevitable that some among the readership will feel that the evidence is one-sided or incorrectly analyzed, or that some small bit of information has been omitted which may lead to an entirely different conclusion of the matter. This is no less the case in the matter at hand, which is perhaps a way of recognizing that it is not so simple a matter as it may initially appear. As the following pages will show, determination of the essential dates of Jesus' life is anything but a simple matter. In addition, the accurate measurement of time has been in fact a rather difficult matter for most of human history, and has therefore attracted the attention of some of the best minds in human history. So my apologies are hereby expressed to any readers who feel that their viewpoint has been slighted, undocumented, or misrepresented in this text. At the same time, however, the primary purpose of this text is to present the story of Christian efforts to understand and measure time, and especially to worship the God who created time. While this study is not intended to be exhaustive, it is the intention of this author to give consideration of the variety of attempts to date the life of Jesus, and furthermore to understand and measure time—a variety which is in fact a central part of this very human story.

1

WHEN WAS JESUS REALLY BORN?

One of the first lessons of United States history which has been learned in multiple generations by millions of school children across the wide expanse of the country is that the first American president, George Washington, was born on 22 February 1732. That date is now somewhat easier to remember since the creation of Presidents' Day as a national holiday. Since both Washington and that other most famous of American presidents—Abraham Lincoln—had February birthdays, the US Congress deemed it appropriate to set aside a day to honor the chief executives of our government on the third Monday of the shortest month of the year. Similarly, millions of students in the United Kingdom have been taught to remember the birth date of their dearest queen—Elizabeth I—on 7 September 1533, and her date of death on 24 March 1603.

But are these dates indeed accurate? Astute observers of dates will note that careful biographical essays stipulate that both England and her colonies were using the "Old Style" calendar during the lifetimes of both of these crucially influential individuals. As noted later in chapter six, a reform of the calendar had been proclaimed by Pope Gregory XIII in 1582, but many Protestant nations such as England were slow to adopt it because it was viewed as one more papal attempt to regain control in lands that had rejected papal authority. The Pope's new calendar—surely an improvement on the old Julian calendar from the first century BC—had eliminated ten days from October 1582 in order to bring the calendar into conformity with planet Earth's astronomical relationship with the sun. Furthermore, the Pope's decree had adopted 1 January as the official beginning of the civil year, replacing the former orientation around 25

March. Consequently, a birth certificate for the date of Washington's birth would have stated 22 February 1731, since at that time the English understanding of the year still began with 25 March. Likewise, it is quite accurate to say that Queen Elizabeth I actually died on the last day of 1602.

What do the birth and death dates of an American president and a British queen have to do with a discussion of the birth date of Jesus, let alone the dates of his death and resurrection? As these simple examples show, it was only a few centuries ago that our ancestors lived by a calendar which most of us would find rather strange and somewhat difficult to use today. More importantly, it brings us face-to-face with the small but unavoidable inaccuracies which have plagued most calendars through most of human history.

"What would you say is the calendar date of Jesus' birth?" This question has been posed by this author to groups of listeners in both churches and the academic classroom. Typical answers include 25 December in the year 1 BC, AD 1, or 0 BC/AD. On the other hand, numerous people have heard critiques of the 25 December date, and so instead assert that he was more likely born in January, in the spring, or in early autumn. Some have claimed that the 25 December date was a Catholic invention intended as an overlay for the pagan Roman festival of Saturnalia, and was used by the church to make money from believers who would travel to a holy site for the occasion. Others have flatly stated that the calendar date of Jesus' birth makes absolutely no difference to them, and that it probably had little significance for early generations of Christians.

Given the lack of calendrical specifics in the Biblical narratives of Jesus' birth, it is tempting to go along with at least the last of those statements and simply celebrate the spiritual meaning of Jesus' birth—the fact that God loved sinful humans enough to become one of us in the person of his Son Jesus, and further, to die a sacrificial death for our sins and to rise from the dead in victory over sin and death. Yet that very statement includes recognition of something more than just a spiritual event. "But after all," someone will say, "the death and especially the resurrection of Jesus are ultimately more important than his birth, are they not?" While the apparent answer is "Yes," we are nevertheless in danger of missing the deeper significance of the matter if we view it as merely a spiritual matter and forget the fact that there was a specific moment in human history when God the Son became incarnate in human form.

To put it another way, even though it may be ultimately less important than the spiritual meaning of Jesus' life, there is an unmistakable quality of historicity and chronology in the event of the life of Jesus and how the events of his life were celebrated by his followers—a quality that should not be missed by those who claim to be his followers 2,000 years later. This book is one attempt to understand more accurately one of those historical aspects—the essential chronological dates of his life.

This chapter begins the investigation of the essential dates of Jesus' life by introducing first of all the several matters of debate based on biblical, historical, and astronomical evidence—and where those bits of evidence lead us in determining the most likely date of Jesus' birth. A second section reviews early Christian calculations of the date of Jesus' birth, along with the theological and calendrical understandings which informed their statements. It is hoped, of course, that the reader will not be satisfied with a discussion focused merely on the date of Jesus' birth, and will continue reading later chapters in order to grasp the larger picture—including the date of Jesus' resurrection from the dead and its importance in the historical and calendrical understanding of early Christians. A third section discusses briefly the critique of the 25 December birth date of Jesus and its alleged connection to the Roman festival of Saturnalia. A final section widens the scope by introducing two long-standing concepts of historical interpretation which have informed the understanding of millions of Christians—the *integral age* of the prophets and the *World Week*.

JESUS' BIRTH DATE: A REVIEW OF BIBLICAL, HISTORICAL, AND ASTRONOMICAL DEBATES

From the standpoint of Roman history and biblical studies, there are several prime issues to be engaged when considering the likely birth date of Jesus. One of those prime issues is the ostensibly assumed date of Jesus' birth as implied in the chronology of sixth-century monk Dionysius Exiguus. It was Dionysius who constructed the western (Christian) chronological system—the same dating system in use throughout most of the world today. Although nowhere in his writings does Dionysius forthrightly state his chosen date of Jesus' birth, most scholars who have studied his writings agree that his use of numbers points to 25 December 1 BC—just seven days before the beginning of the year known as AD 1. Dionysius and his chronological work will be discussed in greater detail in chapters four and five. But what can be stated here briefly is that Dionysius' chosen date for Jesus' birth has been widely criticized as being at least a couple years off the mark.

Why? The birth narrative of Jesus in the Gospel of Matthew (2:1–10) describes unmistakably the paranoia of Herod the Great when he heard that Magi had arrived from the east and were inquiring in Jerusalem about a newborn Messiah. After ascertaining from Jewish religious authorities that the Messiah was to be born in Bethlehem in fulfillment of Micah 5:2, he consulted with the Magi and told them to inform him of the infant Messiah's whereabouts after they had completed their mission in finding him. According to most scholars of recent generations who have researched the date of Herod's death, however, all of this was impossible in December of 1 BC because Herod had already died. Flavius Josephus, a Romanized Jew living in the late first century AD, claimed that Herod's death took place thirty-four years after the death of his predecessor Antigonus and thirty-seven years after his royal recognition by the Romans. By several accounts, this date corresponds to 4 BC[1]—thereby leaving Christians with the rather strange and uncomfortable likelihood that Jesus was born a few years *before Christ.*

It should be noted that recent scholarship has pointed rather in the direction of 1 BC as the correct year of Herod's death. This is based on the assertion that Josephus was incorrect in his dating of two crucial events—the death of Herod and the census mentioned by Luke in relation to the governorship of Quirinius. Concerning the date of Herod's death, Steinmann has contended that the lunar eclipse mentioned by Josephus as occurring just before his death—and therefore an ominous sign—is not the partial eclipse of March 4 BC, but rather the full eclipse of January 1 BC. This date actually fits

---

[1] Josephus' account is found in *Ant.* XVII, 8. This has been confirmed by several as corresponding to 4 BC, although recent scholarship has revived the debate and suggested 1 BC as the true date. See Emil Schürer, *A History of the Jewish People in the Time of Jesus Christ* (New York: Scribner, 1896); Timothy David Barnes, "The Date of Herod's Death," *JTS* 19 (1968): 204–219; P. M. Bernegger, "The Affirmation of Herod's Death in 4 BC," *JTS* 34 (1983): 526–531; Harold W. Hoehner, "The Date of the Death of Herod the Great," *CKC*, 111.

better with Josephus' claim that Herod was nearly 70 years old at the end of his life. Furthermore, an eclipse in January leaves sufficient space for the several events Josephus describes between the eclipse and Herod's death shortly before Passover, whereas an eclipse in March makes for an extremely compressed set of events.[2] Even with an assumption of Herod's death in March of 1 BC, however, Dionysius' apparent assertion of Jesus' birth in December of 1 BC is nonetheless an impossibility.

A second biblical/historical debate has centered on Luke's assertion (2:2) that the census which brought Joseph and the pregnant Mary to Bethlehem just before the birth of Jesus was "the first registration when Quirinius was governor of Syria." It is well-known from other sources that a census was ordered by Caesar Augustus in 8 BC, although its execution in outlying and rebellious provinces like Palestine was inevitably delayed by a few years. One problematic matter is the identification of Quirinius as governor of Syria. Although he was indeed governor of Syria by the time of the next census in AD 6, Quinctilius Varus was the governor in 6–4 BC and therefore possibly at the time of Jesus' birth. Consequently, it comes as no surprise to many that Josephus associated the census rebellion led by Judas the Galilean with the census in AD 6.[3]

Several possible solutions to this difficulty have been offered. Authors such as Crossan and Watts who distance themselves from commitment to historical reliability, let alone inerrancy of the biblical text, have asserted that Luke was simply wrong in his mention of Quirinius, or that he confused Quirinius with Quinctilius Varus.[4] Other solutions more palatable to Bible-believing Christians include the assertion by both Steinmann and Rhoads that it was Josephus rather than Luke who was inaccurate in his dating of the census, and

---

[2] Andrew E. Steinmann, *From Abraham to Paul: A Biblical Chronology* (Saint Louis: Concordia Publishing House, 2011), 230–238. Finegan agreed with the 1 BC date of Herod's death in the revised edition of *Handbook of Biblical Chronology* (Peabody, MA: Hendrickson, 1998), 291–301. The relevant passages in Josephus are *Ant.* XVII.6.1, 4; 9.3.

[3] Josephus' three accounts of Judas the Galilean are found in 1) *Ant.* XVII.6 .2–4, 2) *Ant.* XVII 10.4–8, and 3) *Ant.* XVIII.1.1 [also in *J. W.* II.8.1].

[4] Geza Vermes, *The Nativity: History and Legend* (New York: Doubleday, 2007), 95ff; John Dominic Crossan and Richard G. Watts, *Who is Jesus: Answers to Your Questions about the Historical Jesus* (Philadelphia: Westminster John Knox Press, 1999), 17ff.

that it actually took place about 3 BC. This is based on examination of the three accounts of Judas the Galilean given by Josephus: At first glance Josephus appears to be describing three separate events, but closer examination reveals rather awkward overlap of information and confusion of details of people and events associated with the rebellion. It is therefore quite possible that Josephus, in his effort to present a comprehensive account, misread his sources and presented a single event as multiple events.[5] Also satisfying and defensible to some is the contention of Heichelheim that the Greek term πρώτη in Luke 2:2 should be translated "before" rather than "first"—meaning that Luke was pointing readers toward the census taken before the governorship of Quirinius (but after the proclamation in 8 BC) rather than the later census taken during his governorship (AD 6).[6]

Still another possibility is that the census was implemented in two separate stages: The first stage entailed included registration of all property ownership, which implies that Joseph held title to some land in the vicinity of Bethlehem and therefore would have been required to travel there with the pregnant Mary—thereby providing the conditions of Jesus' birth. By this theory, the second stage involving tax payments took place several years later, providing the occasion of the rebellion.[7] While this approach does not sufficiently resolve all the questions regarding Josephus' use of historical details, it removes the apparent conflict of Josephus with Luke.

Coming to Luke's defense is perhaps eminently sensible in light of the fact that the named recipient of his Gospel was "Theophilus"—who, if in fact he was a flesh-and-blood person, was ostensibly an educated Roman of some political standing who may have known very well the Roman history in Palestine and was therefore acquainted with the governorship of Quirinius. In addition, when one considers other evidences in Luke's account showing that he was

---

[5] Steinmann, 241–249; John Rhoads, "Josephus Misdated the Census of Quirinius," *JETS* 54, No. 1 (March 2011): 69ff. For a critique of Josephus' use of his sources, see Rhoads, 67–68, 85–86.

[6] Fritz M. Heichelheim, "Roman Syria," *An Economic Survey of Ancient Rome* (Tenney Frank, ed. New York: Pageant Books, 1959), 161. Steinmann (241), on the other hand, argues that this is an unnatural reading of the Greek text.

[7] This is the approach taken by Joseph Ratzinger (Pope Benedict XVI) in *Jesus of Nazareth: The Infancy Narratives* (New York: Image, 2012), 61–63), following the suggestion of Stöger (*Lukasevangelium* [London: Burns and Oates, 1969], 172f.).

well-acquainted with the persons and geographic distribution of the Roman governmental system, it is almost inconceivable that he would have made a chronological error of a decade or more in the date of the census or confused one census for another—especially since most Jews in Palestine would remember well the census rebellion under the aforementioned Judas the Galilean and when it occurred.[8] Finally, other suggestions have included the possibility that Quirinius may have held a short-term, temporary governorship over Syria between those of Sentius Saturninus (9–6 BC) and Quinctilius Varus (6–4 BC), or that the purported governorship of Quirinius cited by Luke was actually a reference to his military command over the area at that time.[9]

A third biblical/historical difficulty faces the determination of the specific date of Jesus' birth. Some have asserted that Jesus' birth could not have occurred in December on account of the biblical claim that shepherds were tending their flocks in the fields nearby (Luke 2:8)—a supposed impossibility in December. Bethlehem's geographical location at 31° North Latitude, however, means that the winters are relatively mild, and that sheep can forage virtually year-round.[10] How astute of Luke, moreover, to note that the first announcement of the birth of the Messiah—the one who would become "the Lamb of God who takes away the sin of the world" (John 1:29)—was made to shepherds whose sheep were very possibly destined for sacrifice at the temple in Jerusalem just a few miles away.

Fourthly, biblical/historical debate has likewise swirled around the relationship between Jesus' birth date and that of his slightly older relative John the Baptist, whose father Zechariah was a priest in the order of Abijah (Luke 1:5). Based on the succession of priestly divisions given in 1 Chronicles 24, some have asserted that John's birth took place in mid-April and that Jesus was therefore born in

---

[8] See Mark D. Smith, "Of Jesus and Quirinius," *Catholic Biblical Quarterly* 62, No. 2 (April 2000): 283. Incredibly, however, Smith claims that the Herod of Luke's account is not Herod the Great but his son Archelaus.

[9] See Lily Ross Taylor, "Quirinius and the Census of Judea," *AJP* 54, No. 2 (1933): 120–133; John Thorley, "When was Jesus Born?," *Greece and Rome*, Second Series, 28, No. 1 (April 1981): 81–89.

[10] Paul L. Maier, *In the Fullness of Time: A Historian Looks at Christmas, Easter, and the Early Church* (Grand Rapids: Kregel, 1998), 28; Edwin D. Freed, *The Stories of Jesus' Birth: A Critical Introduction* (Edinburgh: T & T Clark, 2004), 136.

mid-October—very possibly on the Day of Atonement. Yet others have argued on the basis of the re-ordering of priestly divisions recorded in Nehemiah 12 that Jesus' birth is more accurately dated in mid- to late December. The multiplicity of possible dates for Zacharias' priestly service in Jerusalem, however, renders calendrical calculations based on this event extremely tenuous, if not invalid.[11]

Looking to the heavens, astronomical data regarding possible birth dates for Jesus have engaged the attention of astronomers for the past several centuries. Rather fortuitously, it seems that the several years surrounding the birth date of Christ—from the imperial proclamation of the Roman census in 8 BC to the purported Dionysian date for Jesus' birth in 1 BC—were rich in astronomical phenomena.[12] The lunar eclipse mentioned by Josephus as occurring near the death of Herod was later calculated by German astronomer Johannes Kepler as having occurred in 6 BC, which Kepler then concluded was the birth date of Christ.[13] As mentioned previously, however, there was also a partial lunar eclipse visible in Palestine in

---

[11] See "When, exactly, was Jesus Christ born?," http://www.oodegr.com/english/protestantism/imerom_genn1.htm; Steinmann, 252 n 409.

[12] See Jim A. Cornwell, "The Star of Bethlehem and Coma Berenices," http://www.mazzaroth.com/ChapterSix/TheStarOfBethlehem.htm. Rick Larson's "The Star of Bethlehem" (http://www.bethlehemstar.net) is an excellent presentation, although he makes the unnecessary conclusion that Jesus could not have been born in 4 BC or earlier, since—by his calculation—the only notable planetary alignment which would have attracted the attention of astronomers in Babylon ("Magi") occurred in 3/2 BC, and that therefore the Magi arrived in Bethlehem in December 2 BC. He apparently fails to recognize that if the Magi arrived in Bethlehem in December 2 BC, as he claims, it is possible that Jesus was born up to two years earlier—as shown by Herod's command that boys up to two years of age be slaughtered. It is nevertheless the conclusion of this author that Larson's calculation of the date of the arrival of the Magi is roughly correct, although his assertion concerning the planetary alignments is not a necessary supporting argument.

[13] Max Caspar, *Kepler* (New York: Dover, 1993), 227–228. According to Kepler, the star of Bethlehem which drew the attention of the Magi was a supernova, not a planetary conjunction (156). Kepler's *De vero Anno, quo aeternus Dei Filius humanam naturam in Utero benedictae Virginis Mariae assumpsit* was published in Strasbourg in 1613. Shortly after its publication, he was enlisted by Holy Roman Emperor Matthias to give a mathematician's evaluation concerning the adoption of the Gregorian calendar, which had been in use in Catholic dominions already since 1582. Although Kepler (1571–1630) was in favor of its adoption, he demurred from giving strong public support because of the conviction of many Protestant electors that it was merely another attempt at re-imposing papal authority upon their territories.

March 4 BC and another full eclipse in January 1 BC, both of which have been linked with the death of Herod by other researchers.[14] In addition, the appearance of Halley's comet in 12 BC[15], and planetary alignments of Mars and Jupiter in 5 BC, or of Jupiter, Venus, and Mercury with the constellation Leo in 3/2 BC, could easily account for the phenomenon known in Matthew's Gospel (2:2, 9) as the star which guided the Magi to Bethlehem. Consequently, modern astronomers have proposed several viable dates for the birth of Jesus within this timeframe.[16]

JESUS' BIRTH DATE: EARLY CHRISTIAN CALCULATIONS

Where does all of this lead us in determining the birth date of Jesus, or at least a possible range of dates? Although these matters will undoubtedly attract the attention of researchers from a variety of

---

[14] Steinmann (230) asserts that the lunar eclipse of 10 January, 1 BC is the best fit for the lunar eclipse identified by Josephus as occurring just before the death of Herod.

[15] Both Jerry Vardaman and Nikos Kokkinos have identified the appearance of Halley's comet in 12 BC as the star of Bethlehem, although on different grounds. Vardaman, based on his reading of datable passages in Josephus and microletters inscribed on Roman coins of the era, asserted that Jesus' ministry took place about AD 15–19, and that his death occurred in AD 21 at the latest ("Jesus Life: A New Chronology," *CKC*, 55–82). Kokkinos, by contrast, asserted that Jesus' crucifixion occurred in AD 36, based on his dating of the death of John the Baptist and its connection to the events surrounding the marriage of Herod Antipas to Herodias. Accordingly, Kokkinos also interpreted Jesus' statement in John 2:19 in relation to the Jews' designation of the age of the temple as identifying his own age at forty-six ("Crucifixion in AD 36," *CKC*, 133–163).

[16] See Finegan, 307–320; Thorley, 83, 86; Konradin Ferrari-D'Occhieppo, "The Star of the Magi and Babylonian Astronomy,' *CKC*, 41–53. Although his excellent insights are clouded by his rambling and multi-faceted presentation, John Charles Webb (http://www.templeofsolomon.org) presents an interesting case for Jesus' birth as 2 March 5 BC. Similarly, Bruce A. Killian tries perhaps too hard to prove from both astronomical and biblical perspectives to prove that 25 December 1 BC—the ostensibly assumed date of Dionysius—is the accurate date, although his arguments are substantial (http://www.tckillian.com/bible /DionysiusExiguus.htm).

In addition, it should be noted that John Chrysostom (347–407) critiqued the possibility of any usual astronomical phenomenon fitting the biblical description of the star of Bethlehem: ". . . there is not any star that moves by this way, but whether it be the sun you mention, or the moon, or all the other stars, we see them going from east to west; but this wafted from north to south . . . ." ("Homily VI. Matt. 2:1–2," *NPNF*1, X, 37).

disciplines ad infinitum, the present state of scholarship leads us to locate the birth of Jesus somewhere in 3/2 BC. Consequently, it is the opinion of this author that, if it is possible to suggest a likely date of Jesus' birth based on biblical, historical, astronomical, and calendrical evidence, that date is 25 December, 3 BC. The astute reader will note that this choice is in disagreement with that of the aforementioned Dionysius, whose choice of date for Jesus' birth is the beginning of our modern dating system and one of the major bases of this investigation. This choice of date is not intended to imply a level of absolute historical certitude, as if some incontrovertible piece of archaeological evidence had been unearthed in a remote place which precluded all other possibilities. It is merely to assert that this is where our knowledge of the evidence leads us.

If this book were only about the birth date of Jesus, and only about the biblical, historical, and astronomical input concerning his birth, we could leave the matter as it is. But this is equally an investigation into early Christian views of the essential dates of Jesus life. And it is remarkable to note that although it appears to have taken several generations for Christians to take up this question, their calculations were often nonetheless very close to those offered by the biblical scholarship available today.

One of the first Christians to write about matters chronological was Irenaeus (ca. 125–202), bishop of Lugdunum (Lyon). In his *Against Heresies* (ca. 180) he stated that "our Lord was born about the forty-first year of the reign of Augustus," which corresponds roughly to 3 BC. Writing only slightly later, Tertullian (ca. 160–220) asserted that Christ was born in the forty-first year of Augustus and twenty-eight years after the death of Cleopatra. Similarly, Clement of Alexandria (ca. 150–215) stated that Christ was born in the twenty-eighth year of Augustus, "when the first census was ordered to be taken." Then he went on to state that Christ was born 194 years before the death of Commodus (AD 192).[17] Again, this corresponds roughly to 3/2 BC.

[17] Irenaeus, *Adv. Haer.* III.21.3 (*ANF* I, 452); Tertullian, *An Answer to the Jews*, 8 (*ANF* III, 159); Clement of Alexandria, *Stromata* I.21 (*ANF* II, 333). It should be noted that, since Clement links the census to the twenty-eighth year of Augustus, he is referring to the deaths of Antony and Cleopatra and the end of all rival claims to the throne. Clement is therefore not in conflict with the forty-first year of Augustus' reign mentioned by Irenaeus and Tertullian.

Sextus Julius Africanus (ca. 140–220) wrote the five-volume *Chronographiai*, of which we have only fragments remaining in the works of other authors. The *Chronographiai* covered a presumed period of 5,723 years from creation to his own date of writing (AD 221). Africanus dated the passion of Christ as exactly 5,531 years after creation. While Africanus did not give a specific age of Christ at his death and resurrection, by his own numbers 5,531 years after creation yields approximately AD 29, meaning that his earliest possible date for the birth of Jesus is about 4 BC. In an earlier passage, however, Africanus stated that Christ was born 5,500 years after creation.[18] Similarly, Africanus' contemporary Hippolytus (ca. 170–236)—bishop of Portus, near Rome—asserted in his *Commentary on Daniel* that Christ was born 5,500 years after creation, and added that he "suffered in the thirty-third year."[19]

An equally informative source of information attributed to Hippolytus is the set of inscriptions in stone on his sarcophagus, which is located at the Vatican Museum in Rome. The figure representing Hippolytus sits atop a chair, on whose sides are inscribed not only a list of texts attributed to him, but a Passover calendar and an Easter calendar which is presumably based upon it. The Easter calendar on the left of the chair lists dates of Easter for a 112-year period in seven cycles of sixteen years each, covering the years known as AD 222–333.[20] The Passover calendar on the right of the chair likewise gives seven cycles of sixteen years each for the date of Passover, including the day of the week on which Passover was to occur. In addition, Hippolytus included notations beside several of the Passover dates which imply that he believed that the dates for both Passover and Easter would repeat themselves every 112 years (Figure 1).

---

[18] *ANF* VI, 136–138. On the synchronization of AD 221 with *Anno Mundi* 5,723 (years since creation), see Alden A. Mosshammer, *The Easter Computus and the Origins of the Christian Era* (Oxford: University Press, 2008), 387ff.

[19] *ANF* V, 179.

[20] Berard L. Marthaler ("The Date of Easter, Anno Domini, and Other Calendar Considerations: Chronology or Eschatology," *Worship* 73, No. 3 [May 1999]: 199) states that Hippolytus' cycle began with AD 222.

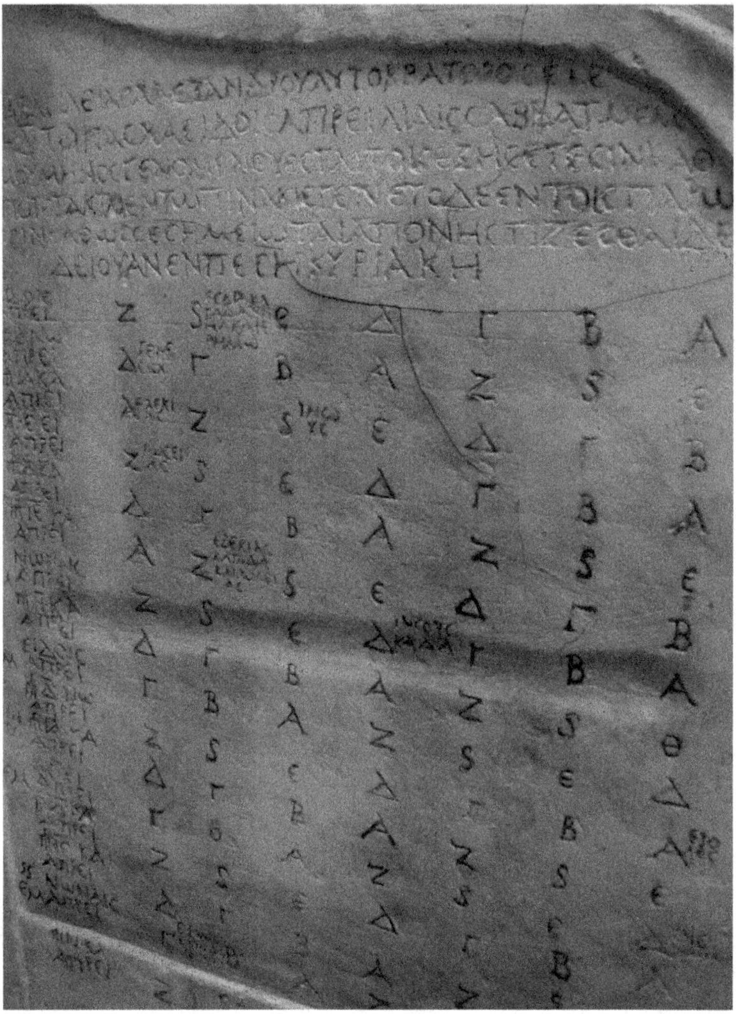

Figure 1. Detail, Passover Calendar, Sarcophagus of Hippolytus (ca. 170–236). Courtesy of the Vatican Museums, Rome. Photo by Author.

With regard to the date of Jesus' birth, an arresting notation on the sarcophagus is included beside the second year of the first cycle of Passover dates, stating, "Genesis Christi" for the Passover date of 2 April (IV Nones April). This could be interpreted as Hippolytus' choice of 2 April as Jesus' birth date, although it is just as likely that he was referring to the incarnation, meaning that he believed that Jesus was conceived in the womb of his mother Mary on 2 April and born approximately nine months later. It is equally important to note, however, that the year for which the notation is apparently given is AD 223. Subtracting two complete 112-year cycles from 223—depending upon which date is chosen as the beginning of a new year—brings us to 2/1 BC.[21]

Still another helpful source of early Christian calculations of the essential dates of Jesus' life is Eusebius (263–339), who became bishop of Caesarea in AD 314, shortly after the accession of Constantine and the consequent legalization of Christianity. His *Ecclesiastical History* states that Christ was born in the forty-second year of the reign of Augustus and the twenty-eighth year after the subjugation of Egypt and the deaths of Antony and Cleopatra. Again, this corresponds to 3/2 BC.[22] Likewise, his *Chronicle* locates the birth of Christ in the third year of the 194th Olympiad, corresponding to 2 BC.[23]

So it is clear that there are multiple ancient sources, both Christian and non-Christian, pointing to 3/2 BC as the likely year of Jesus' birth. But what about the date of 25 December? When it comes to locating a specific day and month for Jesus' birth in early Christian writings, one should be aware that here we enter into the fascinating but risky world of calendrical speculation. For one thing that becomes clear when reading the ancient authors is that they were often attracted to possible connections between dates of significance. Stated another way, one can sometimes detect in their calculations a nearly irresistible urge to correlate events in the life of Jesus with calendrical

---

[21] *PG* X, 875ff. See also the discussion of the sarcophagus in Allen Brent, *Hippolytus and the Roman Church in the Third Century: Communities in Tension before the Emergence of a Monarch-Bishop* (Leiden: E. J. Brill, 1995), 5ff. Unfortunately, modern calendar converters show that Hippolytus' faith in the 112-year cycle was unfounded, as the dates of Passover do not repeat themselves exactly after this interval. 2 April, 2 BC corresponded roughly to 28 Nisan, not the Passover date of 14 Nisan. Similarly, 2 April, 1 BC corresponded to 9 Nisan. The sarcophagus also includes a notation of the supposed date of Christ's Passion, which is discussed in chapter four.

[22] "The Church History of Eusebius," I.5.2; *NPNF*2, 1, 88.

[23] Eusebius' *Chronicle* is known to us only through the reconstructions of Byzantine chronologer George Syncellus, although the tables were preserved and translated into Latin by Jerome, whose work is found at http://www.tertullian.org/fathers/Jerome_chronicle_03_part2.htm. An English translation of the Armenian version of the *Chronicle* is found at http://www.attalus.org/translate/eusebius.html. *OCY* (773–774) and Finegan (284–291) give more comprehensive lists of early Christian calculations, including examples of other years chosen for Jesus' birth, but further show 3/2 BC as the most common.

dates already possessing strong theological meaning. More than mere romanticism, this tendency among early Christians shows that they were looking at this matter with the eyes of faith and seeking to discern the hand of the divine in human history. While this should not deter us from investigating possibilities of a specific day and month of Jesus' birth, it should nevertheless act as a warning sign telling us to proceed with caution.

The earliest reference to a specific day and month of Jesus' birth appears to be that of Clement of Alexandria, who at the same place in his aforementioned *Stromata* stated not only that Christ was born in the twenty-eighth year of Augustus but also that he was born on the Egyptian date of 25 Phacon. Yet he goes on to state that some Christians celebrate Jesus' birth on 24 or 25 Pharmuthi. As noted in the next chapter, the ancient Egyptian calendar was a period of strictly 365 days, with no provision for leap years. As a result, the dates of the Egyptian calendar moved slowly backward through the more widely used Julian calendar of Clement's lifetime, so that 1,461 Egyptian years were equivalent to 1,460 Julian years. Consequently, 25 Phacon in the twenty-eighth year of Augustus corresponded to 20 May 2 BC, although no other early Christian sources have suggested Jesus' birth at this time of the year. Concerning 25 Pharmuthi, if calculating by the Julian calendar, with leap years included, it corresponded to 18 November 3 BC—another date not supported by any other early Christian source. On the other hand, if calculating strictly by the 365-day Egyptian calendar, 25 Pharmuthi corresponded to 6 January 2 BC.[24]

Two strong traditions regarding the birth date of Jesus therefore developed in early Christianity. The earlier of these was the eastern (Greek) tradition which focused on 6 January in the Julian calendar—not because it was 6 January, however, but because that it was assumed to be approximately nine months after the preceding Passover. In other words, it made perfect sense to some early Christians that the life of Christ, the Passover Lamb, began with his incarnation in the womb of his mother Mary on the feast of Passover (14 Nisan 3758 in the Hebrew calendar and 5/6 April 3 BC in the Julian calendar)—with the result that he was born nine months later

[24] Ed Rickard, "The Date of Christ's Birth," Bible Studies at the Moorings, http://themoorings.org/apologetics/chronology/Chrmas.html; *OCY*, 773–774.

on 6 January 2 BC.[25] By the late fourth century, most Christians in the east deferred to their brethren in the west and adopted the date of 25 December for the celebration of Jesus' birth, with the exception of Epiphanius of Salamis (ca. 315–403), who maintained in his *Panarion* that the eighth day of the ides of January (6 January) was the correct date.[26] Even with the gradual shift in favor of the December date, however, the January date retained its festive designation as the time to celebrate both the arrival of the Magi and Jesus' baptism by John.[27]

In the next chapter we will see that the Hebrew calendar, like many other ancient calendars, was luni-solar, meaning that it required the addition of an extra lunar cycle every few years in order to synchronize lunar and solar time and keep the festivals in their proper seasons. Unfortunately, we have no reliable record of when the intercalations of the extra month were made prior to the mid-fourth century AD. This means that speculation concerning possible correlations of dates in the Hebrew calendar with dates in the Julian calendar during the lifetime of Jesus is just that—speculation. While modern calendar converters assuredly bring us very close to the actual ancient dates, the results are still educated guesswork at best.

A second tradition regarding the birth date of Jesus developed in the west favoring 25 December. Like their eastern brethren, western Christians also exercised calendrical speculation, especially with

---

[25] This calculation is amended by modern calendar converters, which show Passover as having occurred on 30 March 3 BC, not 5/6 April. As noted, it should be remembered that we have no record of annual constructions of the Jewish year and its intercalations of extra lunar cycles prior to the mid-fourth century AD. Likely historical re-constructions of both Jewish and Julian calendars are found at http://abdicate.net/cal.aspx, http://www.rosettacalendar.com, and http://www.hebcal .com/converter.htm.

[26] *The Panarion of Epiphanius of Salamis* (Leiden: E. J. Brill, 1987), 201. It should also be noted that Epiphanius was in agreement with the several other early Christian writers already cited who identified the forty-second year of Augustus as the year of Christ's birth.

[27] *OCY*, 21ff. In fact, it was an eastern theologian—Chrysostom—who apparently swayed opinion toward the western date of 25 December for the celebration of Jesus' birth. In his homily "Concerning Blessed Philogonius," dated at 20 December AD 386, he posed Christmas as "the most solemn and awe-inspiring of all feasts" because it preceded the events that form the basis of all other Christological feasts. The fact that he differentiated between the dates of Christ's birth and baptism (Epiphany), shows his support of the western date. See Pauline Allen and Wendy Mayer, *Early Church Fathers: John Chrysostom* (Florence, KY: Routledge, 1999), 191.

dates in the Julian calendar. The earliest known citation of 25 December as the birth date of Jesus is found in the aforementioned *Commentary on Daniel* (4.23) by Hippolytus, where he stated that Christ was born on the eighth day of the kalends of January—which by Roman calculation is 25 December—and that it was a Wednesday. Just over a century later came the *Chronograph of 354*, an almanac of unknown authorship designed for use by Christians around Rome. The notation of "natus Christus in Betleem Iudeae" for VIII Kalends Ianuario is found at the beginning of a list of commemorative dates for Christian martyrs.[28] While these are the earliest known specific links of 25 December with the birth of Christ, in both documents this linkage is presented simply and forthrightly with no emphasis or explanation, which implies that it was already well known and that Christians around Rome had been celebrating Jesus' birth on this date for some time.

But why 25 December? As in the case of 6 January, it is not because of the date itself, but because of its relationship with another date nine months previous—which in this case is 25 March. It was stated perhaps most clearly by Augustine (354–430) of Hippo, who asserted not only that Christ "is believed to have been conceived on the 25th of March," but that it was "upon [this] day he also suffered," and later that "He was born, according to tradition, upon December the 25th."[29] Another example is found in the *Canon* of Hippolytus, where he specifically associated 25 March with the beginning of creation, and in one place even with the date of Christ's birth.[30]

In the minds of many ancient peoples, including the Romans, 25 March was the vernal equinox, and was therefore viewed as the beginning of a new year. In addition, to look into the minds of early

---

[28] See Finegan, 325; Roger Pearse, "The Chronography of 354 AD. Part 12: Commemorations of the Martyrs." *MGH Chronica Minora I* (1892), 71–72, http://www.tertullian.org/fathers/chronography_of_354_12_depositions_martyrs.htm.

[29] *On the Trinity* IV, 5; *NPNF*1 III, 74. The assertion of the *integral age* of Jesus—that he was conceived and died on the same calendar date—will be discussed further in chapter four.

[30] *ANF* V, 179ff. Ogg concludes from Hippolytus' *Canon* and *Chronicle* that he later identified AM 5502 as the year of Christ's birth. It is also surmised that the single occurrence of Hippolytus' association of 25 March with the birth date of Christ may be more properly a reference to the incarnation. See George Ogg, "Hippolytus and the Introduction of the Christian Era," *Vigiliae Christianae* 16, No. 1 (March 1962): 2–18.

Christians and how they processed the question of Jesus' birth date, it is necessary to widen our focus beyond the Christmas event and view it in conjunction with what Christians have always considered to be the most important event in Jesus' life, and indeed in all of human history—his death and resurrection.

When did Jesus rise from the dead? All four Gospels of the New Testament agree that it occurred on the Sunday following Passover—whose date is dependent on the lunar-based Jewish calendar. Since lunar cycles (approximately 29.5 days) and solar cycles (approximately 365.25 days) do not fit together conveniently, the Jews struggled to keep an accurate calendar, as did other ancient civilizations. Like most other calendars of the Mediterranean world, by the first century BC the Jews were utilizing the Metonic cycle—a nineteen-year scheme which synchronized lunar and solar cycles into a workable whole—which is discussed further in the next chapter. This meant that seven out of every nineteen years included a thirteenth lunar cycle (month)—sometimes called an "embolism"—which then also explains why Passover appears to jump all over the calendars of March and April.

Consequently, the variability of Passover's date on the calendars of March and April points to the reason behind Easter's variability of date as well—being ideally the Sunday immediately following Passover. But while this explains the lunar side of the equation in Easter's calendrical calculation, what about the solar side? This brings us back to 25 March. The solar portion of the equation is somewhat simpler, being limited to the single date of 25 March and the astronomical event of the vernal equinox, or the beginning of spring in the northern hemisphere. As explained in chapter four, the eventual Christian formula for determining the date of Easter was that it occurred on the first Sunday after the first full moon after the vernal equinox, which was assumed to be also the Sunday after Passover.

When the Romans began using the Julian calendar in 45 BC, the vernal equinox fell on 25 March. It had been the practice of the Romans and many other civilizations of the ancient world to celebrate the vernal equinox as the beginning of a new year. It was understood as the day of creation, with all things in perfect harmony and balance, just as the earth was when it came fresh from the hands of God.[31] The

[31] Pavel Kuzenkov, "How old is the world? The Byzantine era κατα Ρωμαιους and its rivals," *21st International Congress of Byzantine Studies* (London, 2006): 2–3.

Romans allowed two small errors into the Julian calendar, however—both of which called for later corrections. The well-known error allowed by the Romans was the assumption that the solar year is exactly 365.25 days in length. As we shall see in chapter two, the solar year is slightly short of this figure. This means that over the course of years the solar events which mark our seasons—such as the vernal equinox—slowly move backward to ever earlier dates if the time difference is not corrected. This is why by the fourth century AD and the Council of Nicaea the calculation of the vernal equinox was not at 25 March but 21 March—where it remains today after the Gregorian correction of 1582.

The lesser-known error in the early use of the Julian calendar was that the Romans were initially convinced that a leap year was necessary every third year rather than every fourth year. So twelve leap years were observed between 45 BC (709 AUC) and 12 BC (742 AUC), whereas only nine should have been observed. By this time the error was noticed, and it was proclaimed by Emperor Augustus that the next three leap years—9 BC (745 AUC), 5 BC (749 AUC), and 1 BC (753 AUC)—would be observed as common years. The next leap year was observed in AD 4 (757 AUC). This is an important detail to remember because it affects the day of the week on which any date would fall within in the range of years affected by this adjustment, which happens to include the year of Jesus' birth.

Back to 25 March: Being the anniversary date of the first day of creation, it therefore made eminent sense to many early Christians that God began his re-creation of a redeemed humanity through the incarnation of God the Son into human flesh also on 25 March. Hence, the day on which the angel Gabriel appeared to the Virgin Mary to announce that she would be the mother of the Messiah (Luke 1:26ff) was assumed to be 25 March—still shown on many Christian calendars today as the feast of the Annunciation. Consequently, the traditional western date for Jesus' birth is nine months later on 25 December.

So the date of 25 December 3 BC certainly has some evidences in its favor. But is it possible to go still further, to see any additional calendrical correlations which would add further evidence in favor of this date or any other? While again remembering the caution that comes with entering into calendrical speculation, some possibilities might include the following:

25 December 2 BC also happens to be meaningful in the Hebrew calendar as 2 Tebeth, the last day of the feast of Chanukah, which celebrates the re-dedication of the Temple in Jerusalem after the Maccabean revolt in the second century BC. Looking at it from a Christian viewpoint, how appropriate it is that the birth of the Son of God would take place not only at the approximate date of the winter solstice and at the time when pagan Romans celebrated the birthday of the sun, but also at the culmination of the Jewish festival of light and deliverance.

There are, however, two obvious calendrical difficulties with this date. First of all, this is only a few months before the likely date of Herod's death in March, 1 BC. It therefore makes for an almost impossibly compressed set of several events taking place after Jesus' birth and before Herod's death.[32] In addition, this date makes his conception at Passover unlikely, since Passover in 2 BC was not on or near 25 March, but a few weeks later in mid-April.

Another intriguing possibility is *25 December 5 BC*. That means Jesus' incarnation date was nine months earlier on 25 March, just three days after Passover.[33] In addition, while modern calendar converters show 25 March 5 BC as a Saturday, because of the aforementioned error in the Julian calendar it is likely that it was experienced by people in the Roman world as a Sunday—the same day of the week as the assumed beginning of creation.[34]

---

[32] Luke 2:22ff tells the story of Jesus' presentation at the temple and Mary's purification, which was 40 days after Jesus' birth. The traditional date is 2 February, being 40 days after 25 December. Since Mary and Joseph's offering was a pair of turtledoves—a poor person's offering—the Magi must have arrived later (Steinmann, 252–253). Even if they arrived immediately afterward, this leaves only about six weeks between Jesus' presentation and the lunar eclipse of March, 1 BC which took place shortly before Herod's death. Six weeks is an extremely short timeframe for the arrival and departure of the Magi, the flight into Egypt, and the slaughter of the innocents—all before Herod's death.

[33] This is roughly in agreement with Paul L. Maier ("The Date of the Nativity and the Chronology of Jesus' Life," *CKC*, 127ff) and Harold Hoehner (*Chronological Aspects of the Life of Christ* [Grand Rapids: Zondervan, 1977], 38). In these instances, however, both authors supported the 4 BC date for the death of Herod.

[34] In chapter five it is noted that my calculation is in disagreement with calendar converters available on several web sites. Nevertheless, I will stand by my calculation for the following reason: Curiously, none of the calendar converters accounts for the Roman error of the three-year cycle for the leap year and the years affected by the corrective measures mentioned above. Consequently, every calendar converter shows

Unfortunately, while this date is attractive, it is also too early. Not only is it more than three years before the likely date of Herod's death in March, 1 BC, but it is also before the likely date of the census rebellion in 3 BC and therefore before the census which brought Joseph and the pregnant Mary to Bethlehem. If we assume, however, the consensus argument that Herod died in 4 BC, a 25 December 5 BC birth date for Jesus is possible. Yet it runs into the same problem as that already cited for the date of 25 December 2 BC, in that it makes for an almost impossibly compressed set of events taking place after Jesus' birth and before Herod's death. On the other hand, a birth date for Jesus on 25 December 3 BC faces no such difficulties.

With that in mind, perhaps a small adjustment in the suggested beginning dates of Jesus' life offers an equally meaningful result: By calendrical reconstruction we can estimate that in 3 BC the feast of Passover occurred on 30 March (14 Nisan 3758), which is one full gestation cycle (270 days) from 25 December (18 Tebeth 3759). While perhaps no single set of numbers will meet with complete satisfaction by all interested parties when we are seeking meaningful dates for Jesus' life, we should always remember that we are engaging in calendrical guesswork, and that only God knows the true dates and their significance. At the same time, based on the evidence we have seen from multiple sources, it is the opinion of this author that modern Christians should have no qualms toward celebrating the birth of Jesus on 25 December. For if there is one date that enjoys support from multiple early Christian writers, as well as the several evidences from biblical, historical, and astronomical sources, that date is *25 December 3 BC.*

In chapters four and five we will investigate in similar manner the question of the most likely dates for Jesus' death and resurrection. Again, we will examine biblical and historical evidence. Likewise, we will again engage the chronological statements of early Christians on this matter. More central to that discussion will be the writings of the aforementioned Dionysius Exiguus, who provided the beginning point of our investigation of Jesus' birth date with his ostensibly assumed choice of 25 December 1 BC. Curiously, we have seen little support for his apparent choice in the sources consulted thus far, and therefore in the later chapters we will ask the question of the apparent

---

25 March 5 BC incorrectly as a Saturday, when it appears likely that persons in the Roman world experienced it as a Sunday.

reason behind his choice of dates. Before we leave the present chapter, however, we should briefly engage two matters of historical interpretation which have been debated among Christians and non-Christians alike.

WHAT ABOUT SATURNALIA?

Since at least the late nineteenth century it has been asserted by (especially) liberal and non-Christian writers that Christmas was originally instituted as a counter-festival over against the popular pagan celebrations of the winter solstice, known among the Romans as "Saturnalia" and usually celebrated on 17–23 December. Many have assumed that the choice of 25 December as the birthday celebration of Christ was a political power play by Roman Emperor Constantine I (AD 312–337) and his successors to enforce their new religion on the general populace. Commonly cited in this argument is the fact that Emperor Aurelian (AD 270–275) built a temple in Rome to "Sol Invictus" (the Unconquerable Sun) in AD 274—a half-century before Constantine.[35] This assertion was popularized more recently by Dan Brown's *The Da Vinci Code*.

Unfortunately, this theory possesses some glaring weaknesses. First of all, evidence from patristic sources indicates that Christians were already celebrating the birth of Christ long before Aurelian's construction of the Sol Invictus temple. For instance, the *Constitutions of the Holy Apostles*—a document of unknown authorship, but which originated from Asia Minor in the late second to early third century—lists the birth of Christ along with several other feasts observed by Christians.[36] In addition, even the casual observer can see that the dates of Saturnalia do not exactly match that of Christmas, meaning that it is calendrically impossible to assert so simply that Christmas was merely a Christian overlay intended to replace a pagan festival.

---

[35] For instance, see Stefan Heid, "The Romanness of Roman Christianity," *A Companion to Roman Religion* (Jörg Rüpke, ed. Malden, MA: Blackwell, 2007), 417; "Saturnalia: The Carnaval Celebration that Became Christmas and New Year's Eve," http://www.carnaval.com/saturnalia/. See also the review of literature and arguments by Anthony McRoy, "Christmas: Pagan Festival or Christian Celebration?", http://www.answering-islam.org/pagan/christmas.html.

[36] "Constitutions of the Holy Apostles," V, III, *ANF* VII, 443.

More evidence from patristic sources indicates that Christians were generally offended by the level of feasting, revelry, and sexual license which accompanied Saturnalia. For ancient Christians, celebrating the birth of Christ often meant fasting rather than feasting.[37] In addition, while Saturnalia may have been observed by numerous tribal groups before the lifetime of Jesus, it appears that Aurelian was the first to give it official Roman recognition. While the death of Aurelian likely prevented his persecution of Christians from advancing very far, his attempt to give political and societal backing to Saturnalia shows nonetheless that he was no friend of Christianity.[38]

In fact, the complete lack of a specific connection between Saturnalia and the date of the celebration of Jesus' birth in the writings of early Christians betrays the probable falsehood of the theory. It is therefore likely that the liberal critique of the origin of the celebrations surrounding the birth of Christ is dependent on a skewed interpretation of historical facts—that in fact the Christian celebration of Jesus' birth predated the official recognition of Saturnalia, not vice-versa. While numismatic evidence of the period appears to affirm the claim that Constantine indeed suppressed the cult of Sol Invictus,[39] the prior popularity of the Christmas celebration means that there was no need for him to attempt a simple Christian replacement of Saturnalia. Hence, it appears that this is at least equally a case of a pagan Roman ruler of the third century trying to counteract the growing popularity of Christianity, as it is a case of a Christian Roman ruler of the fourth century trying to counteract the popularity of a pagan practice.

THE 6,000 YEAR HISTORY OF PLANET EARTH

Although it has not garnered as many headlines in recent years, a widespread Christian belief over the centuries with primary

---

[37] Gregory of Nazianzen, "On the Theophany, or Birthday of Christ," *NPNF* (Second Series. Grand Rapids: Eerdmans, 1989 [1893]), VII, 346.

[38] Both Eusebius (*Church History*, VII, 30) and Lactantius (*Of the Manner in Which the Persecutors Died*, VI) interpreted Aurelian's murder by officers of the Praetorian Guard as divine judgment for persecution of Christians. See *NPNF*2, I, 316; *ANF* VII, 303.

[39] Jonathan Williams, "Religion and Roman Coins," *A Companion to Roman Religion*, 160–161 (Figure 11.38).

connections to chronology and eschatology is known as the *World Week*. According to this historical understanding, the six creative days of Genesis chapter one are analogous to the years of human history. Using the popular day-year method of historical interpretation of Biblical texts, it is assumed that all of human history is to take place within a span of 6,000 years—reflective of Peter's comment that "[w]ith the Lord one day is as a thousand years, and a thousand years as one day" (2 Peter 3:8).

The roots of this kind of historical understanding go back at least as far as the *Book of Jubilees*, a Jewish apocryphal text of the second century BC which posited a "Jubilee of Jubilees," or fifty cycles of forty-nine years (2,450 years) between creation and the giving of the Law to Moses. The Jubilee itself, of course, is built on the concept of compounding the number seven—the number of creation. The same text likewise contains allusions to the Messiah reigning on Earth for 1,000 years.[40]

The best-known example of this use of chronology was constructed by Anglican Archbishop James Ussher, which asserted that creation began on October 23, 4004 BC, and that the birth of Christ was 4,000 years later in 4 BC. Ussher's chronology was popularized by its inclusion in the margins of Authorized (King James) Version Bibles beginning about 1650. Quite understandably, Ussher's chronology was later subjected to secular ridicule with the development of "Old Earth" geological calculations. It was given renewed attention and ridicule again in the late twentieth century when it became apparent that his implication of the return of Christ occurring about AD 1996 failed to come to fruition.

Before we completely dismiss Ussher's work as an example of antiquated chronology and an embarrassment to modern Christians, however, two things should be remembered: First of all, he was not, as some have implied, a simple-minded literalist of the Biblical text who merely added the years between the "begats" to come up with a convenient date for creation. On the contrary, if anything, his careful use of the chronologies of other ancient civilizations shows their weaknesses and limitations as well. Secondly, we should remember that Ussher's assumption of a 6,000-year history of planet Earth was

[40] James Barr, *Biblical Chronology: Legend or Science?* The Ethel M. Wood Lecture 1987. Delivered at the Senate House, University of London, 4 March 1987 (London: University of London, 1987), 9ff.

nearly universal among educated persons of just a few centuries ago. A century before the publication of Ussher's chronology, for instance, Martin Luther began his lectures on Genesis by asserting that, "We know from Moses that the world was not in existence before 6,000 years ago."[41]

More importantly for this discussion, it should be noted that the 6,000-year assumption enjoyed wide popularity with early Christian writers as well. As noted already, both Africanus and Hippolytus asserted that Christ was born at roughly 5500 *Anno Mundi*, or since the beginning of the world. Africanus also dated the passion of Christ as exactly 5,531 years after creation.[42] The occurrence of Jesus' birth on or near AM 5500 meant that he was born at the eleventh hour of human history—near the end of the divine plan. Accordingly, Hippolytus went on to assert that the 6,000 years would be followed by a "Sabbath rest." While the consequently implied date for the anticipated return of Christ (approximately AD 500, or AM 6000) was safely beyond the lifetimes of Africanus and Hippolytus, one can only imagine the consternation it likely caused for readers living a few centuries later.[43]

---

[41] *LW* (Jaroslav Pelikan, ed. Saint Louis: Concordia, 1958), I, 3. Duncan Steel's defense of Ussher is very accurate and even admirable, given his admittedly atheistic viewpoint. See *Marking Time: The Epic Quest to Invent the Perfect Calendar* (New York: John Wiley, 2000), 211ff.

[42] The particular date of 25 March for Christ's Passion has been inferred from the fact that Africanus adds no months or days to the number of years separating the Passion from the Creation, whereas he is elsewhere very particular about counting days, months, and weeks (in one place even to the bewilderment of the brilliant sixteenth-century chronologer Joseph Scaliger). Africanus also states, in agreement with some readings of John's chronology, that "the passion of our Saviour falls on the day before the Passover." See *ANF* VI, 136–138; Ogg, 13. A worthwhile critique of the sources and biases of Christian chronologists such as Africanus, Eusebius, and Syncellus is given by William Adler, *Time Immemorial: Archaic History and Its Sources in Christian Chronography from Julius Africanus to George Syncellus* (Washington: Dumbarton Oaks, 1989), 30ff, 232ff. Alden Mosshammer (385ff) asserts that Africanus is actually the originator of the concept of the "Christian Era" attributed to Dionysius Exiguus, who is discussed in chapter five.

[43] A variation on the 6,000-year theme is seen in Augustine, who in a sermon on John 2:1–11 posited that human history is comprised of six ages—the first "from Adam to Noah; the second, from Noah to Abraham; . . . the third, from Abraham to David; the fourth, from David to the carrying away into Babylon; the fifth, from the carrying away into Babylon to John the Baptist; the sixth, from John the Baptist to the end of the world." See "Tractate IX," *NPNF*1, VII, 65. The World Week is also apparent in his understanding of chronology in *City of God* XXII, 30 (*NPNF*1, II, 511). Perhaps

A somewhat related understanding of human history which has been very popular among evangelical Protestants for most of the past two centuries, albeit with its own unique theological agenda, is that found in dispensational theology. Although it asserts no less than eight dispensations, or ages in human history, the seventh takes place during the Millennium and the eighth during "eternity future," meaning that only the first six dispensations are to be enacted during the course of history as we know it. Moreover, in dispensational charts of the divine redemptive plan one encounters the assumption of a 6,000-year limit for human history.[44]

Finally, it is worth noting that not everyone in this modern world, and perhaps especially among Christians, has bought into the Darwinian notion that the cosmos is billions of years in age. The *Creation Museum* in Petersburg, Kentucky, opened in 2007 amid widespread controversy in the media, is an attempt by one Christian organization to give scientific and academic credibility to the theory of a "young" Earth, again with an estimated age of roughly 6,000 years.[45]

It is not the intent of this author to argue either for or against the young Earth theory, or to attempt to define a "Christian" opinion on the matter. The reader with interest in this matter is invited to peruse the several excellent texts on the topic. What should not be missed, however, is the long history of Christians investigating and writing upon this and related topics, and the eschatological expectation implied in nearly every discussion among Christians regarding the age of the Earth and its relationship to the divine redemptive plan.

We will return in a later chapter to eschatological questions, their historic popularity among Christians, and their place today, even after the two-thousandth anniversary of Jesus' birth. In the next few

---

aware of his nearer proximity to Africanus' predicted end to human history, Augustine also advised against speculation and fear-mongering of every kind, reminding his hearers of Jesus' statement in Matthew 24 that "It is not for you to know the times, which the Father hath put in His own power." See *City of God* XVIII, 53; *NPNF*1, II, 394.

[44] A good example of dispensational ordering of human history is found in Clarence Larkin, *Dispensational Truth, or God's Plan and Purpose in the Ages.* (Santa Fe: Sun Publishing, 1998 [1918]).

[45] See Larry Pierce, "The World: Born in 4004 BC?," http://www.answersingenesis.org/articles/am/v1/n1/world-born-4004-bc. For Christian debates on the age of the Earth, see *Three Views on Creation and Evolution* (J. P. Moreland and John Mark Reynolds, eds. Grand Rapids: Zondervan, 1999).

chapters, however, our attention will be focused on time itself and its importance for our ancestors, especially our Christian ancestors. The reasons why time measurement was such a long and laborious task are the subject matter of the next chapter.

2

Down to Egypt, on to Rome

Timekeeping in the Ancient World

Abraham went down to Egypt to sojourn there, for the famine was severe in the land.

—Genesis 12:10

What did the writer of Genesis—most likely Moses—mean by stating that "Abraham went *down* to Egypt"? Many of us moderns will observe on a map that Egypt is somewhat south of Canaan where Abraham had been living, and therefore conclude that Abraham went down because he went south. But we should remember that our assumption of south being "down" and north being "up" is the mental invention of modern maps. Abraham possessed no modern maps, however, so for him "down" meant literally down, as in elevation. The more important matter for Abraham, however, was his own survival against a famine. So he, his family, and his servants left the higher elevations of Canaan and went down to the relatively famine-proof delta of the Nile River.

Most of us reading a text such as this in the twenty-first century have lives that are far removed from the periodic risk of famine faced by our ancestors of several thousand years ago. As nearly as we can tell, it was in such places as that same Nile River Valley of Egypt, and perhaps concurrently in the Tigris and Euphrates valleys of Mesopotamia, that the basics of agriculture were first developed. It was also in these places of early agricultural advancement that our ancestors refined their concepts of time and history or, in other words, learned to measure time so that they would have a history. The more these societies advanced in their practice of agriculture, the more necessary it became to construct an accurate understanding of time and its passage. They simply could not afford to not know when to plant and when to harvest their crops, for the lack of a stable food supply meant nothing less than starvation. So even those groups of people at the fringes of civilization and those more isolated found it necessary to track the passage of time. For all humans, however advanced and sophisticated or backward and benighted, understanding the passage of time was a matter of basic survival.

Speaking of survival, the item that was certainly first and foremost on their minds when they woke up each morning was that of food. While we moderns simply pour a bowl of cereal from a box in our cabinet, or put a frozen pastry in the toaster or microwave, or grab a bite at the deli on the way to the office and the allegedly more important matters of our day, our ancestors spent much of their day in the activities of procuring or growing enough food to sustain themselves to the next. And while many of them were quite successful at cultivating vegetables, fruits, and grains from the fields and forests of their habitat, and hunting or domesticating the animals of those same forests and fields, or catching fish from the oceans and rivers which surrounded them, obtaining food was a never-ending, full-time job. Without our modern luxuries of food preservation, almost anything they gathered or cooked over a fire would spoil within a few days.

Except for those living in the lush tropical forests of our planet's central latitudes, the diets of our earliest ancestors were determined by the annual feast-and-famine cycle of nature's production. The sparse diet of the winter months was limited to the few vegetables that could be preserved through drying, freezing, or salting, along with a few items baked from grains and the meat from the latest kill. For some, winter was literally a "starving time" if weather patterns had hindered the growth of fruits and grain-bearing plants or if viral illnesses or drought had decimated local animal populations during the previous year. Despite the lower level of physical activity during the colder months, humans do not possess the ability to hibernate, and it was therefore necessary for our ancestors to continue hunting and gathering what was available during the winter.

AGRICULTURE AND TIMEKEEPING

In a world in which so many of us obtain our food already neatly harvested and packaged at a store or served on a tray ready to eat, and are therefore not directly involved in the production or processing of our own food, the importance of simple agriculture for our ancestors of several thousand years ago might easily escape us. As they increasingly comprehended the astronomical cycles of the seasons, as well as the natural cycles of the plants on which they and most animals depended for survival, their comprehension was followed closely by manipulation. Since no one had yet developed writing, there are no written records of the agricultural revolution. Yet this is the first revolution—which then made writing and every other aspect of human civilization possible.

We can only imagine the transformation which took place among our ancient ancestors when they began to realize the benefits of planting whole fields of crops rather than ranging far and wide to find them. Likewise, we can only imagine the transformation when they figured out how to keep certain animals in herds and in relatively confined spaces. Once the cultivation of crops and herding of animals had been regularized, our ancestors lived an increasingly settled lifestyle, and the human population began to explode. The safety and security of permanent settlements resulted in a higher survival rate for both children and mothers. This had several important results which provided the beginnings of what is usually defined as human civilization. The first of these was the never-ending need to use more and more land for agricultural production, to feed the ever-increasing number of people. Secondly, because most farmers produced many more crops and animals than were necessary to feed their own family, not everyone needed to be directly involved in the production of food. And certainly not every child of farmers wanted to continue farming as an adult, so many of them specialized in other occupations and traded their services for the food produced by the farmers.

This led to the development of trade centers in villages and towns, where the greater density of people meant a greater variety and complexity of social relationships. That greater variety and complexity necessitated an ever-growing list of social rules designed to protect lives and property, to keep social order, and to punish infractions of the stated rules. Speaking of social order, the first political systems developed around shared ideas of social order, regulation of trade, religious truths, and the need to protect persons and property. And the earliest writing fragments focused on exactly those matters—regulation of trade, religious concepts, and social order.

We could go on nearly ad infinitum to describe the development of the earliest civilizations and their increasing mastery of agriculture. But whatever their level of sophistication in food production and social organization, the arrival of spring and warmer weather (or dry season as opposed to wet season, or sometimes vice-versa) brought an increase in food-procurement activities. Again, the brute necessity of such things can be easily lost on the majority of us moderns whose daily lives are not consumed by the concerns of food production. As noted already, however, our ancestors could hardly afford to miss the windows of opportunity afforded them by nature, for the alternative was not promising.

In addition, the changes of seasons and the annual cycles of planting and harvest meant that certain points in the year rather naturally became times of celebration and worship of the gods of nature who had allowed our ancestors to survive another winter. These celebrations became more elaborate with increasing cultural development. In time, therefore, the ancient Egyptians developed the festival of Shamo, while the Babylonians celebrated Akitu, and the Chinese thanked their ancestral spirits with New Year traditions which continue to this day. In later generations, the Romans celebrated the arrival of spring by naming the first month of their calendar after Mars (March)—who was not only the god of war but also protector of family and fields—while the Germanic tribes farther north worshiped Öster or Estre, the goddess of dawn and the increasing light of spring (whence we get our terms "east" and "Easter").

But without modern calendars and time-keeping instruments like those on our walls and wrists, how did our earliest ancestors figure out when to anticipate another spring and another growing season? They waited for the weather to warm up, of course! But how many days of cold weather were they to expect in a given year? Certainly not every year had the same exact number of cold days or warm days, of wet days or dry days. Some advance planning and basic knowledge of the lengths of the seasons was therefore necessary for the purposes of constructing shelters and storing away food stuffs and other supplies if they expected to survive each and every winter, regardless of its ferocity.

Fortunately for them, our predecessors learned from a very early date to measure time by the great lights God provided in the sky—the sun during the day and the moon and stars during the night. In terms of daily time-keeping, the sun furnished its own primitive clock. Our ancestors could not help but notice the long westward-pointing shadows produced each morning by sunlight shining on innumerable objects just after breaking over the eastern horizon. For that majority of the human population living north of the Equator, it was noted that the shadows were shortest and pointed due north at mid-day (while those living south of the Equator duly noted the southward-pointing shadow, naturally). Likewise, they noticed that the shadows lengthened and bent eastward as the sun sank westward toward the end of the day. In addition, attentive timekeepers among the ancients quickly noted that the mid-day shadow of the sun grew successively longer from day to day and the days grew apparently shorter as the weather grew cooler and winter approached. Conversely, they rejoiced when the mid-day shadow grew successively shorter and the days longer as the warm weather and growing season returned. The sundial was therefore the first and simplest of time-keeping instruments.

Even more useful in their attempts to measure the passage of multiple days and seasons were their night-time observations of the moon and stars. As noted in the next chapter, while the sun takes an entire cycle of seasons to proceed from its highest mid-day point in the sky to its lowest point and back again (nearly 365.25 days), the moon repeats its phases from new moon to full moon and back again more than twelve times as quickly (just over 29.5 days). It is no wonder, then, that our earliest ancestors began constructing their understanding of the passage of seasons with the concept of the month (*moonth*), or lunar cycle. Hence, it is also no wonder that the earliest calendars of every civilization were lunar-based calendars, for even to the casual ancient observer it was clear that just over twelve lunar cycles equaled a complete passage of the seasons, or a solar year.

According to some anthropologists, some of the earliest evidence of human observance of lunar cycles is found in the 15,000 year-old Cro-Magnon paintings in the Lascaux caves in France. There, beneath a painting of a horse one finds a curving row of twenty-nine black dots, corresponding roughly to the length of the lunar cycle. Nearby is found a line of thirteen such dots which ends at a blank rectangle, perhaps symbolic of the half of the lunar cycle from the full moon to the new moon—when the moon virtually disappears from the sky for a few nights.[1]

Figure 2. Stonehenge, England.
Reproduced by permission of Robert and Mary O'Leary.

More decisive evidence and greater precision in time measurement is seen at a number of Neolithic sites. The best-known of these is Stonehenge, on the Salisbury Plain in the south of England. Begun as a Mesolithic ceremonial site perhaps several thousand years ago, by the end of the third millennium BC local residents had erected thirty huge stones—each just over four meters high, two meters wide, and one meter thick, and each weighing about twenty-five tons—into standing positions with somewhat smaller lintel stones lying horizontally on top. In addition to the obviously difficult engineering feat of transporting the huge stones from quarries miles away and hoisting them into position, the site's true significance is seen every year at the summer solstice and other occasions when the sunrise shines directly through the vertically-aligned gaps between the stones and into the center of the circle. Some have also contended that the regular intervals in the locations of three concentric circles of holes

[1] This interpretation has been advanced by Michael Rappenglück of the University of Munich. See "Calendars Through the Ages," http://www.webexhibits.org /calendars /calendar-ancient.html.

(the "Aubrey holes") show that this site was also used to predict lunar and solar eclipses.[2] Part temple and part observatory, the astronomical accuracy of Stonehenge shows an understanding of the heavens perhaps on par with the Egyptians and Babylonians of the same era.

Less-known but equally as impressive are the archaeological remains of time-keeping instruments found at numerous Native American sites. The best example in North America is found at Cahokia Mounds in Illinois. Located just above the flood plain of the Mississippi River and just downstream from its confluence with the Missouri River, this site was once the home of more than 20,000 people during its height of development in the tenth through thirteenth centuries AD. Cahokia appears to have been the capital city of the Mississippian culture, with numerous associated villages in the Mississippi, Ohio, and Missouri river valleys and trade networks stretching from the Great Lakes to the Gulf of Mexico. Most noticeable at Cahokia is "Monk's Mound"—named after the French Trappist monks[3] who once lived there—an earthen mound completed by the Mississippians in the twelfth century and which rises approximately one hundred feet above the surrounding plain. Archaeological excavations at the site have led to the conclusion that the king lived atop the mound, and that the mound also served as the chief religious center for Mississippian culture. From the top of this mound one can see what their king viewed centuries ago—the numerous other smaller mounds with their deliberately varied shapes and sizes, which served as foundations for homes of nobility, burial sites, and boundary markers.

---

[2] Chris Witcombe, "Earth Mysteries: Archaeoastronomy at Stonehenge," http://www.britannia.com/wonder/thehenge.html. Contrary to popular myth, the Stonehenge site predates the development of the religious cult known as the Druids.

[3] French Trappist monks arrived at this site in 1799, after they were expelled from France during the radical phase of the French Revolution. They conducted mission work here among the Native Americans until they were allowed to return to France following the Napoleonic wars. More information on the history of Cahokia Mounds is found at http://www.cahokiamounds.com/explore.

Figure 3. Monk's Mound. Cahokia Mounds State Park, Illinois. Photo by Author.

Figure 4. Woodhenge. Cahokia Mounds State Park, Illinois. Photo by Author.

Recent excavations at Cahokia Mounds have begun to reveal a previously unimagined high level of astronomical and chronological understanding by the Mississippians. Located about one-half mile west of Monks' Mound is "Woodhenge"—so named for its similarity to Stonehenge in England—a series of five intersecting circles or semi-circles of post holes which were constructed over a period of about 200 years. "Woodhenge III," originally constructed about AD 1100, has been reconstructed using the ancient post holes. It consists of a circle of forty-eight posts and a center post, and is about 410 feet in diameter. The posts of the circle are positioned so that from the central observation post one can observe the changing latitude of the sunrise and sunset throughout the year. From this location, at the spring and autumn equinoctia the sun rises directly over Monks'

Mound to the east, lending the impression that the mound itself is giving birth to the sun.[4]

The most impressive example of astronomical and chronological calculation among Native Americans is that achieved by the Maya. Thriving in the Yucatan peninsula and southward (Southern Mexico, Belize, Guatemala, Honduras, and El Salvador) from the eighth century BC onward, they developed the most highly sophisticated pre-Columbian Native American civilization. Agriculture was a necessary supporting pillar of their success. Yet it appears that overpopulation periodically outstripped the production capability of the areas surrounding their largest political and religious centers, and therefore contributed to periods of decline and relocation. Because of the final cultural decline and decentralization of the Maya in the fourteenth century AD, they were not viewed as a military or political threat by the Europeans who arrived nearly two centuries later. Their formerly spectacular centers of trade and worship—already long in decline and disuse—were therefore viewed as mere curiosities by the conquistadores and left for nineteenth-century adventurers to re-discover and interpret.

Figure 5. Ah Cacau Stela. Tikal, Guatemala. Photo by Author.

[4] Cahokia Mounds State Historic Site, "Information on Woodhenge," http://cahokiamounds.org/explore/cahokia-mounds/woodhenge.

The Mayan genius for chronology is displayed prominently on many stela dedicated to their rulers. A prime example at Tikal, the crown jewel of Mayan culture until its abandonment about AD 900, is the stela dedicated to Ah Cacau—better known to English-speakers as Lord Chocolate. It appears that Ah Cacau was one of the most highly revered Mayan rulers, having presided during a period of political and cultural renewal. The date glyph next to his head gives the date of Baktun 10, Katun 15, Tun 15, or about AD 725.

Perhaps the most outstanding example of Mayan expertise in architecture and astronomy is the Pyramid of Kukulcan at Chichen Itza in the Yucatan peninsula of Mexico. The structure visible today was completed in the tenth century, and was actually built around and atop an earlier smaller structure. Since restoration of the site was begun in the early twentieth century, it has drawn multiple thousands of tourists each year. Tourism reaches its height each year during the vernal and autumnal equinoctia, when the genius of the Maya-Toltec builders is on full display. The pyramid, also known as "El Castillo," the name given by Spanish explorers in the sixteenth century, is oriented nineteen degrees east of due north, which is nearly the same as its latitude north of the Equator and therefore nearly equal to the declination of the Earth. This angle, combined with the huge steps of the pyramid itself, produces undulating shadows on the west side of the north staircase as the sun sinks toward the west late in the day. The shadows gradually descend toward the over-sized serpent heads at the bottom of the staircase, producing a vivid illustration of Kukulcan descending from heaven toward his worshipful subjects.[5]

Moreover, even a cursory examination of the constructive details of this pyramid quickly reveals the astronomical and calendrical knowledge of its builders. The Mayans observed a dual calendar—a 365-day civil calendar and a 260-day ceremonial calendar (roughly equal to the period between conception and childbirth), which coalesce at a common starting point only once every fifty-two years. These numbers were incorporated into the pyramid. Each of the four sides includes ninety-one steps. One additional step common to all sides is included at the top to enter the level of the temple. Adding

[5] For more on the astronomical and physical properties of the Pyramid of Kukulcan, see http://www.world-mysteries.com/chichen_kukulcan.htm#Alignments. The serpent deity Kukulcan is also known as Quetzlcoatl, the name given later by the Aztecs.

these steps equals the number of days in the year (91 x 4 + 1 = 365). In addition, this "step-pyramid" (not a true pyramid with smooth sides like those in Egypt) is constructed with nine levels (terraces). Counting those on each side of the stairs gives eighteen—the number of uinals (what some call the "Mayan month") in a civil year.[6] Nine uinals pass between each equinox. Furthermore, close observation of the sides of the pyramid reveals that there are twenty-six depressed panels on each side of each staircase, making a total of fifty-two depressed panels on each side of the pyramid. Fifty-two is the number of years in the great cycle when the civil and ceremonial calendars coalesce.

Figure 6. El Castillo (Temple of Kukulcan). Chichen Itza, Mexico.
Photo by Author.

The modern western calendar used by most of the world today, and even the Christian calendar which is the focus of this volume, were, however, influenced most directly by the calendars of ancient Israel and Rome. Nevertheless, we should glance at least very briefly at some other calendars which were closely related to them and which exercised some influence upon them—those of ancient Babylon, Egypt, and Greece.

[6] See http://www.mayacalendar.com for more information on the Mayan calendar.

CALENDARS IN ANCIENT BABYLON, EGYPT, AND GREECE

As one of the earliest politically centralized civilizations, Babylon created one of the first widely used calendars in the third millennium BC. Alternating months of twenty-nine and thirty days began with the first new moon after the vernal equinox, which was also likely the time of the spring festival of Akitu. By the mid-first millennium BC they had created a nineteen-year cycle to synchronize lunar and solar occurrences, which included seven intercalations of an extra lunar cycle in every nineteen years. It is also worthy of note that some of the Babylonian names for months of the calendar were later adapted by the Jews during their captivity in the sixth century BC—most notably the first Babylonian month of Nisanu, which became "Nisan" in Hebrew usage.[7]

Coming from the same era is the calendar of ancient Egypt. No one knows exactly when it was first instituted; some say it was as early as 4241 BC, while others have suggested 2782 BC, and even others have suggested 1322 BC. What is known is that it was in use well before the occupation of Egypt by the Macedonians under Alexander in the late fourth century BC. Those specific years have been suggested, moreover, because they are the only dates in the ancient history of Egypt when the official beginning of the Egyptian year—the first day of the month of Thoth—actually occurred on the date when it was supposed to occur, in conjunction with the rising of the star Sirius (Sothis, Polaris) and the summer solstice.[8]

As originally designed, the Egyptian calendar divided the year by the three major weather phenomena: "Inundation," or wet season,

---

[7] Further information about the Babylonian calendar and the festival of Akitu is available in Bonnie Blackburn and Leofranc Holford-Strevens, *The Oxford Companion to the Year* (Oxford: Oxford University Press, 1999), 740ff; and at "The Akitu Festival," http://www.gatewaystobabylon.com/religion/akitu.htm; "The History of Christmas," http://www.holidays.net/christmas/story.htm; Graham Kendrick, "Deck the Halls," http://www.grahamkendrick.co.uk/insight/teaching/alpha_deck .htm; "The Reason for the Season," http://mouth-shut.deviantart.com/journal. Some have also suggested that Akitu was celebrated at the autumnal equinox or at winter solstice rather than the vernal equinox. Curiously, some sites seem to be more interested in de-legitimizing practices surrounding the Christian festival of Christmas than in providing solid historical information concerning Akitu or the Babylonian calendar.

[8] *OCY*, 708–709.

began with the rise of the Nile's floodwaters, which occurred always within a few days of the June solstice. It was followed by the four months of "emergence," or winter, and then "dryness," or summer. The alternating months of twenty-nine and thirty days (later changed to twelve months of thirty days), followed by five extra days at the end of the year, gave the Egyptians at most a year of 365 days. This means that the Egyptian calendar slowly backed up relative to the astronomical seasons, with 1,461 Egyptian years equivalent to 1,460 Julian years.[9] Considering the fact that the Egyptians are likely to be credited with some of our basic mathematical referentia such as the 360° circle, division of the day into twenty-four hours, division of the hour into sixty minutes, and division of the minute into sixty seconds[10], it seems rather remarkable that they adhered so resolutely to this calendar—even though they were aware of the error in not counting leap years, and hence countered all attempted calendrical corrections during the Macedonian era.

Meanwhile, similar achievements were underway across the Mediterranean in ancient Greece. Development of a lunar calendar of alternating twenty-nine-day ("hollow") and thirty-day ("full") months, as well as attempts to synchronize lunar and solar cycles, had drawn the attention of Greek astronomers well before the golden age of the fifth century BC, although the politics of competing city-states prevented the cooperation necessary for the creation of a common calendar. The "octaeteris," or eight-year cycle, was one such early attempt at a luni-solar synchronization. A cursory glance at the numbers involved reveals why this block of time was chosen: Eight solar cycles (solar years) equal roughly 2,922 days. Eight lunar "years" of 354 days (2,832 days) plus three extra lunar cycles of thirty days each (for a total of ninety-nine lunations) equals approximately the same number. Therefore, the addition of three additional thirty-day lunations in every eight-year cycle provided a rough luni-solar synchronization. Problem solved! Well, not quite.

The fact that the octaeteris used approximations of the length of lunar cycles, and therefore included an error of about one and one-

---

9 For more information on the Egyptian calendar, see *OCY*, 708ff; Robert Hannah, *Greek and Roman Calendars: Constructions of Time in the Classical World* (London: Duckworth, 2005), 85ff.

10 Mathematics Magazine, "Questions. Why 360 Degrees? Why Base 60?" http://www.mathematicsmagazine.com/Articles/Base60.php.

half days in every cycle, mandated the creation of a better model of lunar and solar synchronization. This was accomplished with the longer and more accurate luni-solar cycle of Meton in the fifth century BC. Most Athenians of that era reveled in their golden age and watched the progress of the construction of magnificent new temples atop the Acropolis—the same location which attracts thousands of visitors today. Easily missed by casual modern tourists, however, are the equally important sites on the adjoining hills. Meton and his associates utilized the nearby hilltop of Lofos Nymfos (Hill of the Nymphs), which was in near proximity to the Pnyx, the location of the polis, or political assembly.

Even today it is easy to see why this location was used for solar measurements during the daytime and lunar and stellar measurements at night. From the western brow of this hill, the ancient astronomers looked directly West toward Mount Aigaleo, whose ridge is oriented north to south (Figure 7). This allowed them to easily track the position of the sunset along the profile of the mountain through the year, as well as tracking lunar and stellar positions at night.

Figure 7. Lofos Nymfos (Hill of the Nymphs), looking west toward Mount Aigaleo. Athens, Greece. Photo by Author.

Furthermore, when one looks eastward from this same location its utility to the ancient astronomers is again revealed. While modern tourists will be impressed with the clear view of the temples adorning the Acropolis, or perhaps even with the view of the Pnyx in the foreground, the ancient astronomers could not help but notice the equally north-south ridge of Mount Ymittos directly to the east of the city (Figure 8). This afforded them equal precision in noting the positions of the sunrise throughout the year, even while working next to the political, religious, and economic centers of the city.

Figure 8. Lofos Nymfos, looking east toward Mount Ymittos. Athens, Greece. Photo by Author.

The ultimate result of Meton's work was the creation of a nineteen-year cycle which included seven intercalations of an extra lunar cycle of twenty-nine days (or, *embolism*) in years 3, 6, 8, 11, 14, 17, and 19. Quite helpfully, this brought the lunar and solar cycles within a few hours of complete agreement with one another.[11] Although it is likely that Meton was not the first to come up with it, and he was certainly not alone in his work, the "Metonic" cycle has forever after borne his name.

THE JEWISH CALENDAR

It is not known exactly how early the Hebrews established their own calendar, although it is apparent that some form of detailed reckoning of time was in operation before the date of the Exodus, whether it is dated in the fifteenth or thirteenth century BC. In preparation for the first Passover while the Israelites were still in Egypt, God told Moses and Aaron, "This month shall be for you the beginning of months. It

[11] There has been considerable debate over small details of how this synchronization was achieved, whether the extra lunar cycle included twenty-nine or thirty days, and in which years the extra lunar cycles were intercalated. See Hannah, 55ff.

shall be the first month of the year for you" (Exodus 12:2). That month is known as Aviv, or Nisan. Like many other ancient calendars, what began as a lunar calendar eventually became luni-solar, with the regular intercalation of additional lunar cycles to prevent the 355-day lunar calendar from drifting backward against the 365-day solar calendar. And again like some of their ancient counterparts, the prime motivating factor was the celebration of a religious festival at the optimum time.

For the Israelites the prime reason for the synchronization hearkened back to Passover, which was to take place during the first month of their religious calendar (not the civil calendar, whose starting point was six months later). The religious calendar was guided by the vernal equinox, which was not allowed to fall more than halfway through the month of Nisan. But the practical demands of thousands of pilgrims traveling to Jerusalem for the festival also became a key factor. No one wanted large numbers of people prevented from attending the festival because of temporal conditions. So over the years the religious leaders developed a set of qualifications which they sought for deciding whether to add an extra month before the beginning of Nisan. These included natural phenomena such as the ripening of barley and blossoming of springtime fruit, and civil phenomena such as roads and bridges being usable for travel.

Details of the historical development of the Jewish calendar have been debated widely by scholars. What can be said confidently is that up until the third or fourth century AD there was likely no regular pattern for intercalating extra lunar cycles into the year, but that it was done merely on an annual basis by proclamation from Jerusalem.[12] Evolution of the Jewish calendar during the early centuries of the Christian era will be discussed in more detail in chapter four. But one small matter which should be mentioned here is that after the several diasporae of the Jews from the eighth century BC onward, a simple annual announcement from Jerusalem became increasingly inadequate, since many Jews lived too far away from Palestine to get the announcement in time and make the proper arrangements. In addition, although the details are again a matter of debate, it is

---

[12] Ephraim Jehudah Wiesenberg asserts, for instance, that details of calendrical intercalations from one year to the next were "jealously guarded by its experts from outsiders, both Jewish and gentile." See Wiesenberg, "Calendar," *EJ* V, 50.

popularly asserted that Rabbi Hillel II wisely foresaw the dissolution of the Sanhedrin in the fourth century AD, and therefore in AD 359 established the pattern of intercalations in years nearly identical to those practiced for several centuries already by the Greeks in the Metonic cycle.

As it concerns this project, the importance of the Jewish calendar for Christianity and chronology is rather obvious. From the apostolic era onward, Christians have understood that the timing of Jesus' crucifixion and resurrection specifically at the occasion of Passover was a deliberate part of the divine plan, so that their redemptive significance would be unmistakably clear. This will be taken up in more detail in chapter four, where we confront the slow but steady drift of Christian calendrical practices away from their Jewish roots as the proportion of Gentile Christians increased and the percentage of Christians with distinctly Jewish heritage diminished. As we will see, however, the unavoidably Jewish roots of the faith continued to be reflected in the Christian celebration of Easter, with its dependence upon lunar cycles, even after becoming relatively independent of the officially-proclaimed dates for Passover.

THE ROMAN CALENDAR

The most direct ancestor of the modern western calendar is the Roman calendar, as it is still reflected today through the names and lengths of our months, the intercalation of the leap year, and its official beginning on 1 January. It is not necessary here to give a complete history of the evolution of the Roman calendar—a history which has produced its own share of debates.[13] What can be said with assurance, however, is that, like the calendars of other ancient civilizations, the Roman calendar went through several stages of improvement and refinement as the Romans themselves slowly expanded from a city-state to one of the most powerful empires in human history. What began as a lunar calendar became a luni-solar calendar, and then finally a solar calendar with only vestiges of its lunar origins.

The major Roman calendrical revision which set the stage for the modern western calendar was instituted during the administration of Julius Caesar in 46 BC, hence the name "Julian Calendar." It was

[13] See Hannah, 98ff; *OCY*, 669ff.

arguably the most important accomplishment of his administration, and certainly the longest-lasting. Two things made it necessary: First of all, as the Romans slowly extended their military and economic reach over the entire Mediterranean world, they inevitably encountered the multiplicity of calendars of other peoples and the consequent difficulties of trade, taxation, and political administration in a world with no shared understanding of time, let alone a common beginning or end of a given year. Secondly, it appears that prior to Julius the Republican government in Rome had employed an official calendar only loosely, so that many local variations and applications were allowed by local officials.[14]

But the demand for efficient administration of an expanding empire forced the end of localized calendars in regions near Rome, and in regions farther distant such as Palestine it meant the imposition of an imperial calendar which required at least nominal assent alongside the local cultural and religious calendar. Like other ancient calendars, the Julian calendar was about more than simple accuracy in time measurement. It was organized not only around solar and lunar cycles, but also around religious holidays. Known as *fasti*, the lists of special days commemorating deities or cultural heroes of the past determined the days on which official business could be transacted. And again like other ancient societies, the *fasti* performed the important function of allowing Romans to remember and reconstruct their past—connecting myth with history.[15]

Again, without launching into further historical explanations, what can be appreciated about this calendar by moderns is the Roman attempt to get beyond the difficulties of synchronizing solar and lunar cycles. They did it by finally divorcing the month, which was originally a measure of the lunar cycle, from the actual length of the lunar cycle. By pre-determining that each year was to have five months of thirty-one days (January, March, May, July, and October[16]) and seven months of thirty days, the 365-day solar year was conveniently divided into twelve roughly equal periods. This removed

[14] Denis Feeney, *Caesar's Calendar: Ancient Time and the Beginnings of History* (Berkeley: University of California Press, 2007), 25–30, 53–57.

[15] Feeney, 68ff, 169; Hannah, 101–104.

[16] August and December were later given a thirty-first day by subtraction from February—the last month of the year, which was thought to be dominated by evil spirits and therefore a worthy target for reduction in length.

the difficulty of remembering exactly at what point one was in the nineteen-year-long Metonic cycle and therefore whether a particular year was to include twelve or thirteen lunar cycles.

Although it was now freed officially from lunar constraints, the Roman month was still organized around the *kalends, nones,* and *ides*, which originally were designations for the new moon, first quarter, and full moon. The *kalends*, whence we get our term *calendar*, marked the beginning of the month, which in its origin referred to the new moon. The *nones*, or first quarter, was designed to occur a week later, and the *ides*, or full moon at the mid-point of the cycle. Even though the lunar cycle no longer determined the length of the Roman month, however, it was thought that the ides should be near the mid-point of the month. Yet not every month had the same length. So it was determined that in a thirty-one-day month the nones would occur on the seventh and the ides on the fifteenth, but in a thirty-day month the nones would occur on the fifth and the ides on the thirteenth.

Additionally confusing for modern eyes is the Roman system of numbering the days between the monthly high points of kalends, nones, and ides. While the first day of the month was simply called the kalends, the second day of the month was not so simple. Rather than being designated "kalends ii," it was either "iv nones" or "vi nones," depending on whether that month had 30 or 31 days. In other words, 2 March on our calendar would be "vi nones," or the sixth day before nones, and 3 March would be "v nones," or the fifth day before nones (counting downward in inclusive fashion, of course). Since March has thirty-one days and April has only thirty, however, a month later 2 April would be "iv nones."

A final note we should make on the Julian calendar is that although 1 January was affirmed as the official beginning of the calendar year, in practice March was still viewed as the beginning of the year, and it is easy to see why. First of all, the month of March was named for the god Mars—official lord and protector of the city of Rome. In addition, numerous ancient civilizations celebrated the vernal equinox as the beginning of another year because it was assumed to be the day of creation. In other words, the vernal equinox reflects creation as it was at the beginning, fresh from the hands of God—with all things in perfect balance. After all, "equinox" means "equal night," or darkness, which implies equal daytime and light as well. Furthermore, the vernal equinox occurs at the beginning of the season of growth and abundance in the northern hemisphere.

As observed in chapter one, this understanding of the vernal equinox as the beginning date of creation was reflected in the assumption of early Christians such as Hippolytus and Augustine that the incarnation of God the Son into human flesh also took place at this date (25 March by the original Julian calculation), and that his later resurrection from the dead (or perhaps his crucifixion) may have also occurred on this date. Finally, the relationship of the equinox with the Jewish feast of Passover set the stage for centuries of debates among Christians concerning the proper dates for Easter.

In retrospect, the calendrical accomplishments of our ancestors of nearly every ancient civilization should meet with our deepest admiration and respect. The multiple astronomical factors which complicated their task are the focus of the next chapter.

3

The Cosmic Puzzle

Astronomy and Time Measurement

The heavens declare the glory of God,
and the sky above proclaims his handiwork.
Day to day pours forth speech,
and night to night reveals knowledge.
There is no speech, nor are there words,
whose voice is not heard.
Their voice goes out through all the earth,
and their words to the ends of the world.

—Psalm 19:1–4

In a world in which so many of us live in cities, whose streets are lit by bright lamps all night and every night, it is sometimes a enlivening revelation to get away from the cities and lights and to behold the wonder of thousands of stars in a clear night sky. This is the sky that was beheld by our ancient ancestors every night as they looked up above their farms and villages, the sky beheld by Jesus as he fasted and prayed in the wilderness, and of course the sky which inspired David to write the psalm.

As seen in a later chapter, one of the underlying beliefs that drove the Scientific Revolution in the sixteenth, seventeenth, and eighteenth centuries was the belief that God is the Creator of the universe as it is, and that deliberate expansion of our knowledge of the physical universe is therefore also an expansion of our understanding of the creative intent of God. It was stated well in the conviction of German astronomer Johannes Kepler (1571–1630) that God had created the physical world "rationally" and "in one measured harmony"[1] in a way that could be grasped by the human intellect. But we should not forget that one of the perennial scientific questions which drove the research of Kepler and others of his era was that of precision in time measurement.

As observed in chapter two, this matter which appears so simple in the modern world as we glance at the clocks on our walls and the watches on our wrists was a matter of serious and ongoing scholarly research by our ancestors in every advanced ancient civilization. In the chapters which follow it will be seen that time measurement continued to be a matter on the front burner of advanced societies until just a few centuries ago, and some would say it has continued so to the present. Why? As seen in this chapter, when considering the relationship between the Earth, the sun, the moon, and the stars one unavoidable discovery is that there are several factors of imprecision built into time measurement, so that what appears initially to be a simple matter is anything but.

For believers in a supreme and eternal God, however, this brings an interesting question: If God indeed created the universe with such rational order and harmony, as Kepler asserted, why is it that time measurement was such a difficult matter for so long? It might be tempting to view God as a kind of cosmic jokester who designed time measurement as a prank to play upon his subjects, although such an image is hardly consistent with the God who loved us enough to send his only Son as our Redeemer. Perhaps it is more accurate, therefore, to view the physical universe as a kind of cosmic puzzle—created by God and given to humanity as an ongoing challenge and invitation to exploration. And with matters of time measurement in mind, another way of looking at it is to ask, if there is indeed a God who created the physical universe as we know it, would it not be just like him to build such imprecision into the system so that we humans, even though we

---

[1] Johannes Kepler, *Harmonices Mundi* (Linz: Godofredi Tampachii, 1619), 102. Author's translation. Also found at http://archive.org/stream/ioanniskepplerih00kepl#page/n10/mode/1up. This and his earlier work *Astronomia Nova* (1609) set down his groundbreaking theories of planetary motion, which changed forever the understanding of the earth's relationship to the sun and other planets.

are by far the crown of his creation, are forced time and time again to stretch our puny minds to understand a bit more of his mind?

This chapter will consider a few of the basic facts of astronomy as it relates to time measurement and the calendars made by our ancient ancestors. This is not intended to be an exhaustive treatment of the subject by any means, and the reader with deeper interests in this field is invited to peruse the numerous excellent texts and web sites which go far beyond the narrow range of topics and concerns presented herein.[2] The narrowness and brevity of this chapter are deliberate, moreover, so that we may focus on those matters of greatest relevance to the study of early Christianity, the calendar, and the life of Jesus. We shall proceed by considering successively those astronomical factors which engaged our ancestors in their process of learning to measure time, beginning with the moon.

THE LUNAR FACTOR

As noted in chapter two, the first astronomical marker of time measurement which attracted the attention of our earliest ancestors was the moon. Making its circuit through the heavens each night and shining infinitely larger and brighter than any stars, yet never staying quite the same as it drifted from one phase to the next, it became an object of awe and worship for many peoples, second in importance only to the sun. The lunar cycle, or *moonth* was found to be useful for several reasons. First of all, the entire lunar cycle is just over twenty-nine and one-half days in length—a number small enough for easy record-keeping. Secondly, the phases of the lunar cycle occur conveniently about a week apart from one another—again, easy for record-keeping. Thirdly, it was noticed from an early date that just over twelve lunar cycles equaled the length of a solar cycle. Furthermore, a full moon was often a benefactor to farmers at

[2] Treatments accessible to non-specialists include Duncan Steel's *Marking Time: The Epic Quest to Invent the Perfect Calendar* (New York: John Wiley, 2000), especially the appendices. Writing not as an astronomer or academic but as a journalist, however, Steel's presentation suffers somewhat by lacking documentation of sources. See also *Universe: The Definitive Visual Guide* (Martin Rees, ed. London: Dorling Kindersley, 2005). Again, the true compendium in this regard is *OCY*, 661ff, 739ff. A wide list of topics and FAQs regarding astronomical aspects of time is found at the web sites of Royal Observatory Greenwich (http://www.nmm.ac.uk) and the United States Naval Observatory (http://aa.usno.navy.mil).

planting and harvest time, affording enough light for a few nights so that work in the fields could proceed around the clock. So again it is no wonder that the Shang Dynasty in China and the Mauryan rulers in India, as well as the Babylonians, the Egyptians, the Israelites, the Greeks, the Romans, and the Maya all utilized lunar cycles in their earliest calendars.

It did not take long, however, for the ancients to discover at least the first layer of the divinely-declared puzzle of imprecision in time measurement. Today the lunar cycle measures twenty-nine days, twelve hours, forty-four minutes, and nearly three seconds in length. Of course, our ancestors did not possess the technology to measure precisely the forty-four minutes, let alone the three seconds. But the proximity to twenty-nine and one-half days allowed them to create calendars of alternating months of twenty-nine and thirty days whose relative accuracy would require the addition of an extra day only about once every three years (32.69 lunar cycles). Yet keeping up precise lunar calculations and adding the extra day at an appropriate juncture was done inconsistently and imperfectly by even the best of the ancient observers.

Furthermore, as if the imprecision of the lunar cycle of roughly 29.5 days was not enough, it was only compounded when the ancients attempted to synchronize it with the solar cycle of nearly 365.25 days. Let us consider for a moment what they were up against: Let us assume that in a given year the lunar cycle began conveniently with a new moon on the same day that the solar cycle signaled the beginning of a new year. Since twelve lunar cycles of 29.5 days amount to 354, they entered by about eleven and one-quarter days into the next lunar cycle in order to get to the end of the solar cycle, or year. By the end of the second solar cycle, they had included about twenty-two and one-half days of the next lunar cycle. And by the end of the third year, they had used thirty-three and three-quarters days, which actually means an entire additional lunar cycle (a thirteenth month?), plus four and one-quarter days.

Assuming that the ancient astronomers were aiming at greater precision, however, they dared not forget the accumulation of the additional forty-four minutes and three seconds in every lunar cycle. This means that by the end of three years they would have already added an extra day in the lunar cycle over against the solar cycle. Consequently, that means there are actually only three and one-quarter days overage in the lunar cycle (not four and one-quarter) at the end of three years. Yet the timekeepers also dared not forget that shortage as they passed through the second three-year cycle and its own set of mathematical intricacies, by which time they found it necessary to also consider . . .

THE SOLAR FACTOR

The star we know as our sun is and always has been the most powerful of astronomical forces for planet Earth and thus for human life. Even long before we were aware that our resident planet is the one that moves in orbits of the sun and not vice-versa, it was the main determinant for our weather and agricultural seasons, and therefore our sense of cyclical time. Quite understandably, the sun was almost universally worshiped as the highest of all deities by the ancients who were immensely thankful for its daily light as well as its heat. But if we are expecting here a level of perfection and simplicity from which all other things can be measured, we will be sorely disappointed. For the cosmic puzzle in regard to time measurement includes what appears to be divinely-declared imprecision at every possible level and juncture.

As if the imprecision of the lunar cycle at slightly over 29.5 days is not enough, it is here that a second level of imprecision must be added. By modern measurements, the solar cycle (year) is 365 days, five hours, forty-eight minutes, and forty-five seconds in length. It is only eleven minutes and fifteen seconds short of what would be a convenient mathematical figure for calendrical calculation—365.25 days. That convenient figure was the assumption of several of the advanced ancient civilizations, including the Romans when the Julian calendar was instituted in 45 BC. Unfortunately, however, that convenient figure is incorrect and, again, imprecision reigns. So what most of us learned in grade school about the length of the year needs to be taken with a healthy grain of salt if we are to achieve anything approaching true accuracy and precision.

The small shortage of eleven minutes and fifteen seconds might be understood more widely in the current generation if it were not for the infamous centennial and millennial transition through which we so recently passed. A backward glance at the calendar from AD 2000 will confirm that it was a leap year. While most will respond begrudgingly with a "ho hum," since it is commonly taught that every year divisible by four is a leap year, we should not. As seen in chapter six, the Gregorian correction of the calendar in 1582 (in addition to eliminating ten days from October) declared that the centennial years—those years ending in "__00"—were not to be counted as leap years unless they were divisible by 400, not just by four. Why? Because of that shortage of eleven minutes and fifteen seconds, of course, which adds up to about three days and three hours in the course of 400 years.

In other words, the western solar calendar was made significantly more accurate by the simple omission of three leap year days (29 February[3]) in every 400 years, and it was assumed that the easiest place to remember it was at the end of the century, when calendrical consciousness is closer to the forefront of the public mind. And lest it be forgotten, attentive readers will no doubt notice that the remaining shortage of three hours at the end of every 400 years means that an additional leap year day needs to be omitted about every 3,200 years. Long before that, however, the current century will come to its end in AD 2100 and someone will need to remind that generation that it is not a leap year.

The Gregorian calendar and the historical circumstances of its development will be discussed in more detail later, but what should not be missed at this point is what appear to be ever-increasing levels of the cosmic puzzle—that is, ever-increasing levels of imprecision in time measurement. When one considers the challenges which faced the ancients in understanding the lunar cycle, the solar cycle, and how to synchronize the two, it is no wonder that it took them multiple generations and multiple attempts. In addition, it should be remembered that literacy was fairly limited to cultural elites, which is not to presume ignorance upon our ancestors but rather to recognize that opportunities for education and furtherance of knowledge were much more limited. So Kepler's confident declaration about the rationality of the "one measured harmony" of the universe would have been unfortunately lost on many of the ancients, simply because

[3] Calendrical purists will object that the true leap year day is the bissextile (twice-sixth) day, which is 24 February. In other words, vi kalends March in the Julian calendar—the sixth day to the beginning of March, or 24 February—was placed on the calendar twice in a leap year. This is because it was considered the un-luckiest day of the year, being the sixth day from the end of the last and shortest month of the year. In modern practice, however, the leap year day is 29 February. See *OCY*, 92–93, 677ff.

of the size of the numbers required for accurate calculations in time measurement.

Since the solar cycle is generally understood as a year, however, an additional question which should be asked is that of the precise definition of the year. The most common astronomical definition is known as the *Tropical Year*, which is the measure of time required for passage through the seasons from one vernal equinox to the next, whose length at just under 365.25 days has been the focus of discussion.[4] It comes as a surprise to some, however, that the passage of an entire Tropical Year does not quite bring our planet back to the same astronomical point in its orbit around the sun. That measure of time is the *Sidereal Year*, which is roughly twenty minutes and twenty-four seconds longer than the Tropical Year at 365 days, six hours, six minutes, and nine seconds. Astute readers will of course recognize that if the Sidereal Year were the standard measure of time rather than the Tropical Year, we would be slowly adding leap years rather than subtracting.

Other definitions of the year are certainly available. From an astronomical perspective, it might make sense to utilize the *anomalistic* year, which is roughly equal to the Sidereal Year in length but is the measure of time from one perihelion[5] passage to the next. Or the year could be calculated from our relationship to stars other than our sun. Or the civil year could be begun on a date other than 1 January. Or we could return to the practice of many of the ancients and celebrate the vernal equinox as the beginning of our year, and do so more accurately than they. Since, however, the current calendar has enjoyed the benefits of modern time-keeping technology, well-intended suggestions by scientists and international organizations are likely to remain just that.

---

[4] Steel (380–383) tries perhaps too hard to differentiate his self-defined *Vernal Equinox Year* from the Tropical Year and to commend it as the best true measure of the year.

[5] The perihelion is the point in the Earth's rather elliptical orbit at which it is closest to the sun, at just under 90.5 million miles. It occurs each year at about 3–5 January. At the other end of the spectrum is the aphelion, or the point in the orbit at which the Earth is farthest from the sun, at about 95.5 million miles. It occurs each year at about 4–7 July.

THE GEO FACTOR

To a limited extent, it could be said that all of the factors of imprecision in time measurement heretofore discussed—and especially the solar factors—are really *geo*, or Earth factors, since the Earth is by default our observational vantage point. After all, even though the winter solstice, summer solstice, and both equinoctia are all events we observe with reference to the sun's changing position in our sky, they have nothing to do with any movement by the sun itself, but rather with the ever-changing tilt of the Earth with respect to the sun. This movement is best represented by what is known as the *ecliptic*.

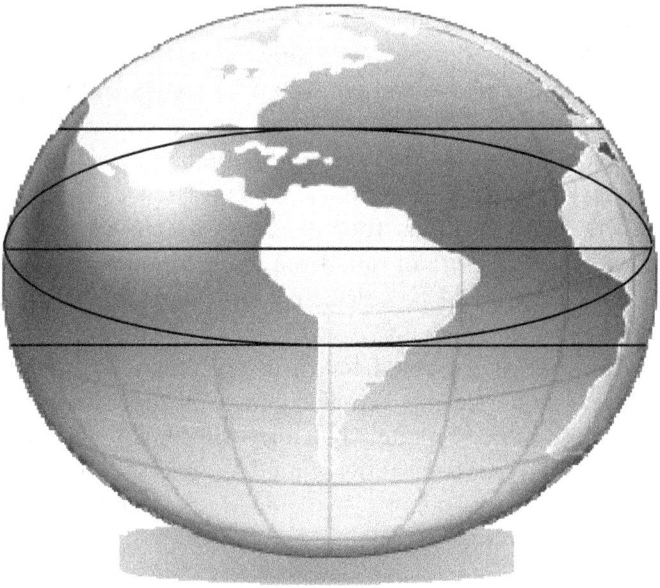

Figure 9. The ecliptic.

The ecliptic is represented by the intersections of the lines in the figure above. The horizontal line through the center of the Earth represents the Equator—the imaginary line which encircles the center of our globe at 0° Latitude. The northernmost ("uppermost") line represents the Tropic of Cancer at roughly 23°, 27' North Latitude. It is the farthest point north at which the sun shines directly overhead, which takes place at the summer solstice for the Northern Hemisphere (winter solstice for the Southern Hemisphere), about 21 June each year. Conversely the southernmost line represents the Tropic of Capricorn at roughly 23°, 27' South Latitude. It is the farthest point south at which the sun shines directly overhead, which takes place at the summer solstice for the Southern Hemisphere (winter solstice for the Northern Hemisphere), about 21 December each year. The arc which begins at the Equator, curves northward to the Tropic of Cancer and back southward to the Equator represents the actual path of the Equator at the December solstice. Conversely, the arc which begins at the Equator, swings southward to the Tropic of Capricorn and back northward to the Equator represents the actual path of the equator at the June solstice.

Stated more accurately, the Earth's tilt, or inclination, on its axis between the poles does not actually change at all during the course of the year, but remains constant at nearly twenty-three and one-half degrees. But the axis itself changes constantly relative to the sun. One way to envision this is to imagine the axis as a long rod that runs through the planet and sticks out at both north and south poles. At the Earth's perihelion in January, the end of the rod sticking out of the north celestial pole points slightly away from the sun at nearly 23.5° away from straight "up." Meanwhile, the opposite end of the rod sticking out of the south celestial pole points slightly toward the sun at 23.5° away from straight "down." This means the sun is shining directly overhead near the Tropic of Capricorn (approximately 23.5° South Latitude)—the beginning of summer in the Southern Hemisphere. Conversely, at the Earth's aphelion in July, the end of the rod sticking out of the south celestial pole is pointed slightly away from the sun at 23.5° away from "down," and the end of the rod at the north celestial pole is pointed slightly toward the sun at 23.5° away from "up"—the beginning of summer in the Northern Hemisphere.

Another way of understanding the ecliptic is viewing it as the intersection of the Earth's *orbital plane* with its *celestial plane*. The orbital plane may be envisioned as a huge flat disc extending outward from the sun to the Earth, with the Earth slowly making its way around the outer edge of the disc at an average distance of about ninety-three million miles. The celestial plane, on the other hand, may be envisioned as another huge flat disc extending outward from the Earth's equator into space. As the earth makes its annual circuit around the disc of the orbital plane, its inclination at 23°, 27' creates the strangely-shaped ecliptic around the Earth—perhaps something like a badly-made bagel, which is widest at its outermost edge, and sharply angled (not rounded) so that it is very thin at the edge of the hole in the center, which represents the Earth.

What does the Earth's motion represented by the ecliptic have to do with time and our perception of time? Clarifying any possible effects that the lack of the ecliptic would have on our gravitational or magnetic fields is best left to geophysicists. From the perspective of time and weather, however, some results would be immediate and obvious to every person living on this planet. Without the Earth's motion represented by the ecliptic, the planet would be virtually stuck at a never-ending equinox. The region of the planet known as the tropics would be a much narrower band, and the sun would forever shine directly overhead only at the equator. The temperate zones both North and South of the equator would expand toward the equator, but shrink farther away from the poles. In addition, the temperate zones would not experience the seasonal changes of weather and temperature known today, and would be somewhat stuck with the kind of weather experienced annually at the vernal and autumnal equinoctia. There would be a never-ending pattern of twelve hours of daylight and twelve hours of darkness, with gradually decreasing daylight and increasing darkness as one moves toward the poles. Consequently, there would be no noticeable growing season or fallow season for crops—meaning that the planet's production capability would be significantly decreased—and no seasons of hibernation or awakening for numerous species of animals. Finally, the polar regions would experience very little light or heat year-round, and would therefore undergo significant expansion into what are today the temperate zones.

Without the changing altitude of the sun in the sky and without noticeable seasonal changes in weather, our ancestors would have had less motivation to track the passage of time, except perhaps to note their own increasing age. They might have still utilized the moon and stars as astronomical markers of time, and might have still connected them with deities worthy of worship, but they would not likely have been connected to agricultural cycles. And without a need to measure the seasons and the sun's changing height in the sky, they might not have developed calendars in the first place. Furthermore, we might wonder whether agriculture—the very foundation of the earliest civilizations—would have developed at all, or at least on any scale approaching the historical reality.

Some readers might perceive beneath these observations a kind of Judeo-Christian apologetic, or "God argument"—an attempt to highlight the unique qualities of our planet and its astronomical location in order to argue for a theistic origin, or what is known popularly as "Intelligent Design." Indeed, in recent years a fair number of scientists and researchers from a variety of disciplines have advanced arguments in favor of the thesis that the best explanation for the structure of the universe, the uniqueness of planet Earth, and the complex nature of biological life on this planet is direct causation by an intelligent Designer, or Creator.[6] Similarly, it might be said that a rather backward God-argument is implied in this discussion behind the details of the lunar, solar, and geo factors of time measurement. Yet it is far from the simplistic and unilateral declaration of a backwoods preacher waving a Bible over his head which is sometimes imagined by agnostic secularists who caricature Christians as uneducated and closed-minded hillbillies standing in the way of civilizational progress.

If the astronomical factors of time measurement say anything about theistic possibilities, they would argue much like the aforementioned Kepler and his colleagues during the Scientific Revolution, who viewed their work as a glimpse into divine mysteries even as they marveled at the heretofore unseen vastness of the universe. In other words, the more we understand about the incredibly complicated and interconnected structure of the cosmos, the more plausible theistic design becomes. And in regard to time measurement, we are brought back to the question asked near the beginning of this chapter: If there is indeed a God who created the physical universe as we know it, would it not be just like him to build such imprecision into the system so that we, even though we are by far the crown of his creation, are forced time and time again to stretch our puny minds to understand a bit more of his mind?

---

[6] See Lee R. Strobel, *The Case for a Creator: A Journalist Investigates Scientific Evidence that Points toward God* (Grand Rapids: Zondervan, 2004); Paul Davies, *The Mind of God* (New York: Touchstone, 1992); Michael Behe, *Darwin's Black Box* (New York: Touchstone, 1996); William A. Dembski, *Intelligent Design: The Bridge Between Science and Theology* (Downers Grove, IL: IVP Academic, 1999). A contra argument is given by Richard Dawkins in *The Blind Watchmaker: Why the Evidence of Evolution Reveals a Universe Without Design* (New York: W. W. Norton, 1996).

In the next chapter we will follow the paths of early Christians who attempted to do just that—to stretch their minds to come up with an accurate calendar. Yet their stretching had an innately theological motivation—the celebration of what they understood as the most important event since the beginning of time—the resurrection of Jesus Christ from the dead.

4

DOCUMENTING THE DIVINE

JESUS' RESURRECTION, EARLY CHRISTIANITY, AND THE EASTER CYCLE

One of the truly wonderful, yet unavoidable and sometimes bewildering facts of modern Christianity is its apparently endless variety. In any sizeable city in those regions of the world where Christianity has achieved social prominence, and perhaps especially in the western world, one encounters congregations with a wide array of theological emphases, social outreaches, positions on political issues, and even some congregations who self-consciously specialize in ministry to a particular ethnic group. In the United States alone there are several thousand ecclesiastical organizations, and multiple thousands of independent congregations claiming no ecclesiastical affiliation whatsoever. Perhaps most noticeably, in Christianity of the twenty-first century one encounters a spectrum of worship styles spanning the entire history of the faith from practices that would be recognizable to ancient Christians all the way to the latest musical genre and technologically sophisticated proclamation.

Given the pluralistic cultural assumptions of the modern western world, this variety in expression is viewed as one of the strengths of the Christian message. But it was not always this way. In fact, during the first few centuries following the lifetime of Jesus, Christians often struggled painfully to come to agreement with one another—not only on the necessary theological issues of the person and work of Christ, but on the proper date for celebrating those events which showcased his unique person and work—his crucifixion and resurrection.

In the modern western ecclesiastical calendar, the celebration of Jesus' resurrection can occur as early as 22 March, and as late as 25 April. But who came up with these official dates for the celebration of Jesus' resurrection from the dead? And how? And why? Is it even possible to determine the actual date of Jesus' resurrection? The first task of this chapter will be exactly that—to examine evidence from biblical and early Christian sources concerning the likely calendar date of Jesus' resurrection from the dead. The second task will be to follow the transmission of that knowledge as it developed into the Easter, or Paschal cycle during the early centuries of the Christian era. In addition, it will be seen that the whys and hows of the Easter cycle are bound up with several important political, theological, and sociological developments of the same period.

WHEN EXACTLY DID JESUS RISE FROM THE DEAD?

From a very early date, Christians did not hesitate to celebrate holy days and seasons, having inherited a rich tradition of annual observances from their already-ancient Jewish heritage. Informed by that heritage, even the first generation of Christians understood the calendrical significance of the event that defined their faith—that the death and resurrection of the Messiah Jesus had taken place precisely at the time of the Passover, so that his role as the redemptive Passover lamb whose blood was shed for the sins of humanity would be unmistakably clear. The precise dating of the event of Jesus' resurrection and its annual celebration from the perspective of secular solar calendars of that era, however, was not uppermost in the minds of early Christians, since most of them shared the heritage of the Jewish Passover celebration already, and since they were occupied at least intermittently with the matter of their own survival in the midst of periodic persecution from both Jewish and especially Roman sources.

In addition, the common belief of the initial generations of Christians in the imminent return of Jesus and the establishment of his earthly kingdom mitigated against an historical consciousness in this-worldly terms, and therefore against the creation of a Christian calendar of annual celebrations. Many Christians of the first generation after the lifetime of Jesus viewed their weekly and even daily gatherings as celebrations of his resurrection. Nevertheless, the matter of dating did become an issue with the passage of time and as the Christian movement spread outward both geographically and socially from its base in Palestinian Judaism. This was especially the case after Christianity was legalized by Emperor Constantine with the Edict of Milan in the early fourth century, and was made the official religion of Rome by his successors near the end of that same century.

JESUS' RESURRECTION DATE: BIBLICAL EVIDENCE

When it comes to the question of dating Jesus' resurrection, the place to begin, naturally, is with the four canonical gospels of Matthew, Mark, Luke, and John. But we must not rush immediately to the Passion narratives if we wish to take into account all the relevant information concerning the day, month, and year of Jesus' death and resurrection. Why? While the gospels do not identify the particular year of Jesus' Passion, Luke tells us (3:1) that John the Baptist began preaching in the desert and baptizing his repentant listeners "in the fifteenth year of the reign of Tiberius Caesar." So the place to begin in dating Jesus' resurrection is with the beginning of his ministry.

The Date and Length of Jesus' Ministry

From the testimony of several Roman sources, we can date the beginning of the reign of Tiberius Caesar to September AD 14, shortly after the death of his predecessor Augustus. This would place the beginning of John's ministry and his baptism of Jesus—in the fifteenth year of Tiberius—probably in the summer of AD 29.[1] Luke also tells us (3:23) that Jesus "was about thirty years of age" when he began his ministry. Irenaeus suggested in the second century that Jesus may have been near forty years of age at the beginning of his ministry, that he had a ministry of perhaps a decade or more, and that he was nearly fifty years old at the time of his death. His interpretation has gained few followers, however, since there is no indication in any of the canonical gospels that Jesus' ministry was more than a few years in length.[2]

---

[1] See the discussion in Finegan, 330ff; Steinmann, 219. This effectively precludes the aforementioned assertion of Vardaman that Jesus' ministry spanned the years AD 15–19 ("Jesus Life: A New Chronology," *CKC*, 55–82).

[2] Irenaeus, *Against Heresies* I.XXII.5–6; *ANF* I, 391–392. This is one of the suppositions of the aforementioned theory by Kokkinos that Jesus' crucifixion occurred in AD 36 ("Crucifixion in AD 36," *CKC*, 133–163). See also the discussion in Steinmann, 257ff.

Luke's use of ωσει ("about") in stating that Jesus was "about thirty years of age" is matched in only one other instance in reference to a person's age—when he states that Jairus, the ruler of a local synagogue, came to Jesus pleading for the life of his only daughter, who was "about [ως] twelve" (8:42). Given the fact that children change markedly in appearance as they pass through childhood and adolescence, Luke's statement of this girl's age likely means that her age was known to be within a year or so on either side of her twelfth birthday. By the same token, his use of a cognate word to describe Jesus' age at the beginning of his ministry likely means that Jesus was within a year or so on either side of his thirtieth birthday.[3] Interestingly, dating the beginning of Jesus' ministry to the summer of AD 29 at the approximate age of thirty also confirms the conclusion in chapter one concerning his likely birth date: If Jesus was born in December 3 BC, his thirtieth birthday occurred in December AD 28. Even while allowing some latitude in possible dates for his birth, this affirms the accuracy of the data we have gained from multiple sources.

The next matter, then, is that of the length of Jesus' public ministry. This question has engendered some debate among scholars in recent years, as the gospels provide two potential answers based on the occurrence(s) of Passover.[4] All three Synoptic Gospels mention only one Passover (Matthew 26:2; Mark 14:1; Luke 22:1)—the occasion associated with Jesus' crucifixion. Meanwhile, John mentions three Passovers (2:13, 6:4, 11:55). The popular options for the length of Jesus' public ministry therefore have been either one year or three years. In actuality, however, this means one and one-half or three and one-half years: Assuming that Jesus' ministry began by the late summer of AD 29, there are far too many events recorded in the Synoptics to fit into the period of just several months between late summer of AD 29 and Passover in early spring of AD 30. A one and one-half year ministry of Jesus, however, dates his crucifixion at Passover in AD 31, which at least allows enough chronological space for the events recorded.[5]

---

[3] This follows the argument of Steinmann, 220 n331.

[4] See the helpful discussions in Finegan, 350–353; Steinmann, 258–264; Hoehner, *Chronological Aspects*, 29–38.

[5] It could possibly be argued that Jesus' baptism by John and the beginning of his public ministry occurred in early autumn, AD 28, which was just after the beginning

There are other indications in the Synoptics, however, in favor of a three and one-half year period for the ministry of Jesus. In other words, all three Synoptics record stories indicating additional seasons of spring. Firstly, all three Synoptics include early in Jesus' Galilean ministry the story of his disciples plucking heads of grain from stalks as they passed through the fields in order to relieve their hunger (Matthew 12:1; Mark 2:23; Luke 6:1), and consequently being critiqued by the Pharisees for ostensibly breaking the Sabbath. This would have happened in the spring, when wheat begins to ripen. Secondly, all four gospels tell the story of Jesus feeding the 5,000 (Matthew 14:13–21; Mark 6:30–44; Luke 9:10–17; John 6:1–15), which is placed near the end of his Galilean ministry. On this occasion, John specifically notes that the Jewish Passover Feast was "at hand" (6:4), and Mark notes that Jesus told the people to sit down on the "green grass" (6:39), which implies that it was springtime.

So it is easily demonstrable that a ministry period of at least two and one-half years is required by the gospels. In addition, three other matters should be noticed which suggest the strong possibility of a fourth Passover during Jesus' ministry—meaning a total period of roughly three and one-half years: Firstly, John's omission of the story of the disciples plucking heads of grain might genuinely imply another springtime in addition to the three associated with his three Passovers. Secondly, the several events recorded in the Synoptics occurring before the story of plucking heads of grain imply a timeline of more than just the several months between Jesus' baptism and the Passover of the next spring. Thirdly, the unnamed "feast" mentioned

---

of the fifteenth year of Tiberius (17 September), which consequently allows for the crucifixion of Jesus at Passover in AD 30—supported by some as the correct year of the crucifixion and resurrection. This would place Jesus' baptism near Yom Kippur—therefore strengthening John's announcement of Jesus as "the Lamb of God who takes away the sin of the world" (John 1:29).

There are two problems with this theory, however. First of all, modern calendar converters date Yom Kippur to 16 September, AD 28, which is just before the beginning of Tiberius' fifteenth year in office. This paints Luke as a sloppy and inaccurate historian. In addition, judging by Roman practices in counting regnal dates, it is somewhat likely that the fifteenth of Tiberius' reign was counted as beginning on 1 January, AD 29. Secondly, it is rather illegitimate to use a statement from John's gospel to support the supposition of a one and one-half year (or even shorter) ministry of Jesus when it is clear that John portrays a scenario of at least two and one-half years, if not longer. The force of John's identification of Jesus as the Lamb of God works just as well if the baptism is placed one year later in late summer, AD 29.

by John (5:1) may possibly indicate another occurrence of Passover. While the question of the strict chronological accuracy of any of the gospels has engendered a discussion of its own among scholars, the fact of their general agreement with one another likely shows that none of them is far away from strict event-by-event accuracy. That being said, the most popular option for understanding the length of Jesus' ministry has been a period of three and one-half years, as seen also in the writings of early Christian leaders. But first we will examine biblical evidence for . . .

Dating the Events of Jesus' Passion

Once again, the gospels present us with a less-than-uniform picture of events, but one that is not without solution when we look carefully beneath the surface. A first question is the chronological relationship between Jesus' crucifixion and the Passover Seder. According to some scholars, Jesus' crucifixion appears to have taken place after the Seder in the Synoptics (Matthew 26:17; Mark 14:12; Luke 22:7), but before the Seder in John (18:28, 19:14).[6] This is a false dichotomy, however, which is resolved when we note other details in John's narrative and observe the bigger picture of the usage of key words and phrases.

At first glance, a reading of John 19:14 ("Now it was the day of Preparation of Passover"[7]) allows the aforementioned implication that Jesus' crucifixion occurred on 13 Nisan—a day of preparation for the Seder meal to take place that evening. But this contradicts the clear indication of John 13:1 which links the evening meal with Jesus'

---

[6] See Joachim Jeremias (*The Eucharistic Words of Jesus* [New York: Scribner, 1955, 177–183) and Annie Jaubert (*La Date de la Céne* [Paris: Gabalda, 1957]), both of whom argue in favor of the Synoptic dating. On the other hand, John A. T. Robinson (*The Priority of John* [London: Meyer Stone, 1985], 147ff) argues for the veracity of John's account. A related debate in these sources is whether the Synoptic authors reflect the Sadducees' dating of Passover (beginning on the evening of 14 Nisan) while John prefers the Pharisees' dating (beginning on the evening of 15 Nisan). See also *OCY*, 722, 729; Greg Killian, "Passover Chronology," *The Watchman*, http://www.betemunah.org/chronology.html. This matter, too, appears to be a false and distracting issue, and is resolved when one sees the essential harmony between John and the Synoptics offered on the following pages. In addition, we shall see in the following section that it is clear that early Christians viewed 14 Nisan as the true date of Passover.

[7] This is clarified in the *New International Version* to state "It was the day of Preparation of Passover Week."

anticipation of his own sacrificial death. The rest of this narrative—known as the "Upper Room Discourse" of chapters fourteen through sixteen—fits exactly the kind of conversation we would expect from Jesus and his disciples when he told them that he was the true Passover Lamb, as stated more directly in the Synoptics (Matthew 26:17–30; Mark 14:12–26; Luke 22:7–38). With no clear indications from John to the contrary, his account certainly appears to describe the Passover Seder Jesus would have shared with his disciples on 14 Nisan, just as we read in the Synoptics.[8]

So how should we understand John's statement? The difficulty is resolved when we observe the larger picture of the phraseology involved. Other uses of "day of preparation" (Matthew 27:62; Mark 15:42; Luke 23:54; John 19:31) imply strongly that this was a colloquial term referring to Friday as the weekly day of preparation for the Sabbath, since physical activities and food preparation were strictly limited on the Sabbath.

Moreover, it should be noted that "Passover" and "the Feast of Unleavened Bread" (the eight-day feast beginning with Passover) were often used interchangeably by the four gospel writers. One of the most notable instances in this regard is that of the boy Jesus accompanying his parents to Jerusalem "for the Feast of the Passover" (Luke 2:41ff). Since it is highly doubtful that common, working-class Jews such as Joseph and Mary would have traveled all the way from Galilee for a festival lasting just one day, it is therefore likely that "the Feast of the Passover" was understood to include the eight-day period. This is also the obvious case in Acts 12:1–4, where it is stated that Herod Antipas ordered both the execution of James and the arrest of Peter "during the Feast of Unleavened Bread," with the intention of putting Peter on trial "after the Passover." Since Passover is at the beginning of the feast and not the end, Luke's mention of Passover at the end of this chronological sequence is incomprehensible if it cannot be substituted for the Feast of Unleavened Bread. In other words, if the two terms were not

[8] For more discussion on whether John depicts the actual Passover meal, see Finegan, 355–358; Steinmann, 275. See also Steinmann's "Table 55. Jesus' Ministry According to the Gospel of John," 261.

interchangeable, this would be a rather awkward and confusing statement by Luke.[9]

It is therefore clear from scripture itself that there is no essential conflict between John and the Synoptic authors over the character and dating of the Last Supper Jesus shared with his disciples. It can be stated confidently that the Last Supper was indeed the Passover Seder. In addition, it can be said that the Last Supper took place on the Passover date of 14 Nisan and on a Thursday evening. The events of Jesus' arrest, trial, crucifixion, death, and burial all took place in the course of the hours stretching from that Thursday evening through the following Friday afternoon, so that Jesus was in the tomb before sundown on Friday. Moreover, it must be remembered that since the Jewish day begins at sundown, all of these events took place within the limits of 14 Nisan—the ultimate affirmation of his divine role as Passover Lamb slain for the sins of humanity. Furthermore, since it is also clear from the gospels that Jesus' followers were anxious to bury him before the beginning of the Sabbath on Friday evening, by counting in inclusive Hebrew fashion we can therefore affirm that he was in the grave for three days.[10] Consequently, Jesus fulfilled his own prophecy (Matthew 12:39–40) that he would be "in the heart of the earth" for three days. Finally, it means that the Hebrew date of Jesus' resurrection from the dead on the following Sunday morning was 16 Nisan.

But where does that place these events in the Roman (Julian) calendar of the period, which is the basis for the modern western (Gregorian) calendar that most of us are accustomed to using most of the time? Calendrically speaking, we are looking for a year in which

---

[9] Curiously, this is not accepted by Finegan (357–358), who has chosen to follow the suggestion of John P. Meier that the Last Supper of Jesus with his disciples was not the Passover meal or some other ritual meal of Judaism, but rather kind of farewell meal deliberately arranged by Jesus because he knew what was about to happen. Consequently, Meier is compelled to claim that the Synoptic authors presented their versions of this meal as if it were the Passover meal when they clearly knew it was not.

[10] For a more detailed chronological account of the week of Jesus' Passion, see Steinmann, 288ff. Rather providentially, we might add, the hurried burial of Jesus necessitated the return of the women to the tomb early on Sunday morning after the Sabbath so that they might anoint his body with oil and spices (Mark 16:1; Luke 23:55–24:1). That necessity provided the opportunity for discovering the empty tomb and the initial appearance of the resurrected Jesus (Matthew 28:8–10; John 20:10–17).

14 Nisan fell on a Friday within the range of AD 26–36—the years during which Pontius Pilate was Roman governor in Judea. Based on the painstaking research of earlier generations of scholars and the confirmation offered by modern calendar converters, it can be stated with confidence that there were only two such occurrences: 7 April AD 30 and 3 April AD 33. As noted previously, we have no reliable record from the first century of which years included the intercalation of an extra lunar cycle, and hence must admit that this is a bit of historical guesswork. Yet since the events surrounding Jesus' crucifixion have attracted so much concentrated attention over the years, what can be said is that multiple generations of scholarship from a variety of disciplines have pointed to one or the other of these two dates as the most likely choice.[11] Furthermore, of these two dates it appears that the second is preferable for two reasons.

First of all, the careful approach of Pontius Pilate in dealing with Jesus and his accusers implies that this event took place sometime after AD 31. Pilate was appointed in AD 26 by Sejanus, the captain of the Praetorian Guard who was allowed by Tiberius to act as prime minister for a number of years. Luke tells us (13:1) of an incident preceding the ministry of Jesus concerning some "Galileans whose blood Pilate had mingled with their sacrifices"—which would fit perfectly with the anti-Semitic attitude of Sejanus who, among other things, had apparently persuaded Tiberius to expel all Jews from Rome in AD 19. But in October AD 31, Sejanus was executed by order of Tiberius because of plots against the royal family, and the anti-Semitic policies were thereafter reversed.[12] This gives insight into Pilate's obviously ambivalent attitude toward Jesus and his accusers: Luke tells us (23:20) that he wanted to release Jesus, and John affirms (19:12) that he "tried to release him." But the Jewish leaders countered with, "If you release this man, you are not Caesar's friend." Had this occurred before Sejanus' fall from power, Pilate could have afforded to ignore their implied threat of sending complaints to Rome and asserted his own will in releasing Jesus. But his accession to their demand shows that the new and more accommodating policy toward Jews was already in place.

---

[11] See Finegan, 359–362, 366–369.

[12] Paul L. Maier, "Sejanus, Pilate, and the Date of the Crucifixion," *Church History* 37 (1968): 8–10; Finegan, 362; Steinmann, 282–285.

Secondly, we are looking for a year in which a lunar eclipse was visible from Jerusalem on 14 Nisan. Why a lunar eclipse? None of the gospels mentions a lunar eclipse, although all three Synoptic authors state that "there was darkness over all the land" roughly between the sixth and ninth hours of the day, or from noon until mid-afternoon (Matthew 27:45; Mark 15:33; Luke 23:44). While this initially appears to describe instead a solar eclipse, it should be noted that a solar eclipse is astronomically possible only at the new moon, not during a full moon as is always the case at Passover.[13] It is therefore more likely that a strong windstorm in the Arabian desert had stirred up enough dust to obscure the sunlight and give the appearance of dusk at mid-day.

But why a lunar eclipse? Peter's sermon several weeks later at Pentecost included a quotation of Joel 2:28–32, citing not only the prophet's statement concerning the outpouring of the Holy Spirit, but also the accompanying astronomical signs. In other words, it appears that Peter interpreted Joel's prophecy that "the sun shall be turned to darkness and the moon to blood" (Acts 2:20) as being just as literal in the divine plan as "I will pour out my Spirit on all flesh" (2:17). And immediately after this citation he proceeded to remind his listeners of the concluding events of Jesus' life. While it may have been ultimately a small point in Peter's sermon, his hearers would certainly have remembered the mid-day darkness of the day of Jesus' crucifixion.

But what about the moon turning to blood? Ancient Babylonian and Chinese astronomical annals record only one full lunar eclipse observable from Jerusalem at Passover in the period of AD 26–36, and that is on 3 April AD 33. Furthermore, a blood red moon is actually a good description of the moon's appearance as it slowly rises out of the shadow of the earth and is seen against the backdrop of a sky which is still clouded with billions of particles of dust. It is estimated that in the late afternoon and early evening of 3 April, the moon began to rise in the sky over Jerusalem with 20% of its disc still in the umbra (complete shadow) of the earth, and the remainder of its disc in the penumbra (partial shadow). The effect would have been quite stunning, with the umbral portion of the disc colored deep red and the penumbral portion an orange-yellow. Since the sun was only

[13] This follows the discussion of Colin J. Humphreys and W. G. Waddington, "Astronomy and the Date of the Crucifixion," *CKC*, 172–179.

beginning to set at this time, the large amount of light remaining in the sky would have provided a brilliant background highlighting the moon's appearance.[14] Again, while this may have been a small point in the larger context of Peter's sermon, his appeal to astronomical signs in the process of interpreting the spiritual manifestations at Pentecost and ultimately in his proclamation of Jesus as the Messiah further validated his message and moved many of his listeners to faith in Jesus on that occasion.

Consequently, it is the opinion of this writer that, based on biblical and historical evidence, the most likely date of Jesus' crucifixion is Friday, 3 April AD 33 on the Julian calendar (1 April by Gregorian calculation), which corresponds to the Passover date of 14 Nisan 3793 on the Hebrew calendar. By extension, this means that the most likely date of the most important event in human history—Jesus' resurrection from the dead—is Sunday, 5 April or 16 Nisan of the same year.

JESUS' RESURRECTION DATE: EARLY CHRISTIAN EVIDENCE

Given the gravity of the events to which Jesus' followers were eyewitnesses, it is perhaps remarkable that they left any chronological clues at all. Yet they did, as seen already, and certainly the most important of these clues was the fact of Jesus' crucifixion at Passover, so that his role as "the Lamb of God who takes away the sin of the world" (John 1:29) would be unmistakable. That understanding is reflected in the writings of early Christian leaders—many of the same writers who commented on the date of Jesus' birth—even as they struggled to coordinate the lunar-based date of Passover with the more widely understood solar-based Roman (Julian) calendar.

Again, the earliest known reference in this regard comes from Clement of Alexandria in his *Stromata*, where he comments on the practices of the followers of Basilides: "And treating of His passion, with very great accuracy, some say that it took place in the sixteenth year of Tiberius, on the twenty-fifth of Phamenoth; and others the twenty-fifth of Pharmuthi, and others say that on the nineteenth of Pharmuthi the Saviour suffered."[15] The chronological clues given here point to the crucifixion of Jesus in AD 29 or 30, and on the

[14] Ibid., 176–179; Steinmann, 285–287.

[15] *Stromata* I.21; *ANF* II, 333. Basilides was a "theologian of Gnostic tendencies" who was prominent in Egypt in the mid-second century (*ODCC*, 141).

Julian dates of 21 March (25 Phamenoth in the Egyptian calendar), 14 April (19 Pharmuthi), or 20 April (25 Pharmuthi). Equally telling is Clement's assertion in the same section that "it was necessary for Him (Jesus) to preach only a year." Given Clement's lack of critique of these dates, it appears that he agrees with the followers of Basilides that these are at least possible dates for Jesus' crucifixion. Unfortunately, of the six possible dates cited (three each in AD 29 and AD 30), only one of them falls on a Friday—14 April AD 30—and even that date does not correspond to the most likely Passover date for that year (7 April). Furthermore, as seen already, it appears most likely from the gospels themselves that Jesus' ministry lasted decidedly longer than one year, and that the better choice for his date of crucifixion and resurrection is AD 33.

While all of Clement's suggested dates now appear to be objectionable because of information which was perhaps not available to him, his suggestion of 25 Phamenoth and of Jesus' age at the crucifixion may have nevertheless provided the precedent followed by other Christian writers soon thereafter. For example, in his *Answer to the Jews* Tertullian stated that Christ was "about [thirty] years of age when he suffered," and further that it was "in the consulate of Rubellius Geminus and Fufius Geminus, in the month of March, at the times of the passover, on the eighth day of the calends of April, on the first day of unleavened bread."[16] In other words, Tertullian pointed to 25 March as the date of Jesus' crucifixion. While the Egyptian month of Phamenoth in those years was roughly equal to the Julian period of 25 February-26 March, if Tertullian and other early Christian writers assumed a simple correspondence between the months of Phamenoth and March they would have been doubly attracted to the twenty-fifth day of the month for the aforementioned reason that 25 March was the original Julian date of the vernal equinox and was therefore viewed as the anniversary date of creation. Unfortunately, while 25 March AD 29 was a Friday, it appears that Passover was nearly three weeks later.

Nevertheless, support for the AD 29 date is found in Hippolytus' *Chronicle*, where he equated his own year of writing (AD 234/235) with year 5738 since Adam, the year of Jesus' birth with year 5502 from Adam, and the year of Jesus' crucifixion with year 5532 from

[16] *An Answer to the Jews* 8; *ANF* III, 160.

Adam. As observed in chapter one, year 5502 from Adam is roughly 2 BC, which means that year 5532 is AD 29.[17] Likewise, close observation of the aforementioned Passover cycle inscribed on the side of his sarcophagus reveals that beside the last date of the second sixteen-year cycle are the words ΠΑΘΟΣ ΧΣ (pathos Christos), or "passion of Christ." This inscription is exactly thirty years past the inscription noted in chapter one regarding the birth of Christ, which showed Hippolytus' belief that Jesus was born in the year known as 2 BC. Similarly, this notation denoting thirty years later is again equivalent to AD 29. The inscribed date of Passover for that year is 25 March, and the Easter date inscribed on the opposite side is 27 March.[18] As noted in chapter one, however, elsewhere in his *Commentary on Daniel* Hippolytus stated that Christ suffered "in his thirty-third year."[19]

Augustine cites the date of AD 29 in *City of God*, asserting that "Christ died when the Gemini were consuls, on the eighth day before the kalends of April."[20] A slightly different dating is found, however, in Jerome's Latin version of Eusebius' *Chronicle*, stating that Jesus "came to the Passion in the eighteenth year of Tiberius," or the third year of the 202nd Olympiad (AD 31/32).[21]

The early Christian fixation on 25 March is even more evident in the chronology of Alexandrian monk Annianus, writing in the early fifth century. Thanks to the work of ninth-century Byzantine historian George Syncellus, significant portions of his chronology have been preserved. Annianus' calculation of the ages of the earth differed by several years from that of Africanus, with the result that his chosen year of Christ's birth is by our counting AD 9. He was also quite possibly the first author to calculate the future dates of Easter by employing a 532-year master calendar cycle, which multiplied the number of days in the week (7) by the number of years in the leap-

---

[17] Finegan, 364–365.

[18] Also interesting is the simple inscription of ΙΗΣΟΥΣ for three years later—equivalent to AD 256, which is two complete 112-year cycles past AD 32. An unanswered question is whether it might be Hippolytus' way of suggesting this as an alternative date for Jesus' crucifixion and resurrection.

[19] *ANF* V, 179.

[20] *City of God* XVIII.54; *NPNF*1, II, 395.

[21] Roger Pearse, "Jerome's Latin translation of Eusebius' *Chronicle*, Part 2, http://www.tertullian.org/fathers/jerome_chronicle_03_part2.htm.

year cycle (4), and multiplied that result (28) by the number of years in the Metonic lunar cycle (19). This combination of multiplications made it possible to track any particular date of the year regardless of its day in the week or position in the lunar cycle. This was all the more advantageous for Annianus because Alexandrian astronomers had meticulously tracked the lunar cycles for centuries, with the result that they were recognized as the prime authority in astronomical and calendrical matters in the late Roman world.

Although Annianus may have been simply repeating the statements of earlier chronologers such as Africanus—many of whose works are no longer extant—or spelling out their implication, he is certainly one of the first Christian writers to fit together the mystical connections of the chronological puzzle, hence identifying 25 March not only as the date of the vernal equinox but also as 1) the date of creation, 2) the date of the incarnation of the Son of God, and 3) the date of Jesus' resurrection from the dead. Consequently, he also pointed to 25 December as the date of Jesus' birth (nine months after the incarnation on 25 March) and to 23 March as the date of his crucifixion. In addition, as the result of his multiple computations he identified AM 5534 (AD 42) as the year of Jesus' Passion because of the agreement of all the solar and lunar data: 25 March AD 42 was a Sunday just after the full moon (22 March) which was the day immediately following Passover (15 Nisan)—which was in agreement with his reading of John's gospel.[22]

Unfortunately, Annianus' chosen dates for Jesus' death and resurrection occurred not during the reign of Tiberius (AD 14–37) as asserted in Luke, but during the reign of Claudius (AD 41–54). Furthermore, it appears at first glance that searching the ancient calendars for an occurrence of 25 March in combination with the Hebrew date of 16 Nisan on a Sunday which could possibly be the

---

[22] Hebrew calendar converters are found at http://www.hebcal.com/converter, http://www.abdicate.net/cal.aspx, http://www.rosettacalendar.com, and http://www.chabad.org/calendar/1000year_cdo/aid/6225/jewish/Date-Converter.html. Additional discussion of Annianus and his bearing on later Byzantine chronologies is found in Georges DeClercq, *Anno Domini: The Origins of the Christian Era* (Turnhout: Brepols, 2000), 29ff. A worthwhile critique of the sources and biases of Christian chronologists such as Africanus, Eusebius, and Syncellus is given by William Adler, 30ff, 232ff. Mosshammer asserts that Africanus is actually the originator of the concept of the "Christian Era" attributed to Dionysius Exiguus, who is discussed in the next chapter. See *The Easter Computus and the Origins of the Christian Era* (Oxford: Oxford, 2008), 385ff.

date of Jesus' resurrection within the widest possible range of years (AD 27–34) is nothing more than a search in vain. In close proximity is 25 March AD 31—the ostensibly assumed date of Jesus' resurrection by Dionysius Exiguus (discussed in the next chapter)—which appears to have occurred on 13 Nisan, at least a few days too early. Also in close proximity are 28 March AD 23 (which corresponds to 17 Nisan, but at least a few years too early), 24 March AD 26 (corresponding to 16 Nisan, but also too early), 25 March AD 34 (corresponding to 17 Nisan, but probably too late and on a Thursday), and 24 March AD 37 (corresponding to 18 Nisan, but again too late). In other words, a problem to be recognized by modern researchers when reading Annianus and some other early Christian writers is the fact that there does not appear to be a single possibility of locating the resurrection of Jesus on 25 March in any of the years in which this event actually could have taken place.[23]

This matter will be taken up again in the next chapter from another angle when considering the work of Dionysius Exiguus and his apparent calendrical choices. What should perhaps be stated at this point, however, is that this is not meant to call into question the value of the work of Annianus or any of the other early Christian chronologers. It is merely to point out the numerous and unavoidable calendrical difficulties which have faced Christian chronologers since the beginning when it comes to determining the essential calendar dates for Jesus' life.[24]

The Integral Age of Jesus

One other matter of interest should be noted here, as referenced in chapter one—the so-called *integral age* of Jesus. Although not referenced specifically in Scripture, it was a Jewish mystical belief that the lives of true prophets were given divine validation by

---

[23] An interesting approach is taken by the author of *The Shepherd's Page* which leaves Jesus inexplicably in the grave for five days (counting inclusively). Using only clues from Scripture, the Jewish calendar, and astronomy, he makes the case for Jesus' death having taken place on Wednesday 9 April AD 27 (14 Nisan 3787 in the Jewish calendar) and his resurrection on Sunday 13 April (18 Nisan). See "The Dates of My Lord Jesus Christ the Messiah," http://www.abdicate.net/dates.aspx. By inclusive Hebrew counting, this approach leaves Jesus in the grave five days.

[24] For a discussion of still other dates suggested by early Christian writers, although none with the widespread and long-lasting popularity enjoyed by 25 March, see *OCY*, 775–776.

beginning and ending on the same calendar day. Going back at least as far as the *Book of Jubilees* in the second century BC, it appears to have been a popular idea among early generations of Christians as well. It is implicit in the writings of such leaders as the aforementioned Clement of Alexandria, Sextus Julius Africanus, and Hippolytus.[25]

As seen in chapter one, it was stated clearly by Augustine in *On the Trinity* (ca. AD 415) when explaining the significance of numbers and calendrical dates in the life of Christ. He asserted not only that Christ "is believed to have been conceived on the 25th of March," but that it was "upon [this] day he also suffered."[26] And we have just seen the equal significance this concept held for the calculations of Annianus.

Unfortunately, we have likewise seen the extreme difficulties faced by early Christian chronologers when attempting to match dates for the beginning and end of Jesus' life which fulfill all the necessary calendrical requirements—requirements such as the day of the week, the day of the Julian solar calendar, the day of the Hebrew lunar calendar, and all occurring in years which fit the historical circumstances describing the beginning and end of Jesus' earthly life. For instance, in chapter one it was observed that it is possible that Jesus' conception in the womb of his mother Mary (the Incarnation) occurred on the Passover date of 30 March 3 BC, and that he was born approximately nine months later on 25 December 3 BC. By the same token, it would be equally possible that the incarnation took place five days earlier on 25 March. But 25 March 3 BC was not the date of Passover.

Similarly, Augustine's enthusiasm in asserting that Jesus was not only conceived on 25 March, but that his later crucifixion was on that date as well, is certainly understandable. Yet while 25 March was on a Sunday in AD 29, it does not appear to have been the Passover. Moreover, we have just established that Annianus' suggestion that Jesus' resurrection occurred on 25 March AD 42 simply cannot be

---

[25] James Barr, *Biblical Chronology: Legend or Science?* The Ethel M. Wood Lecture 1987. Delivered at the Senate House, University of London, 4 March 1987 (London: University of London, 1987), 9ff; Thomas C. Schmidt, "Hippolytus and the Original Date of Christmas," http://chronicon.net/blog/chronology/hippolytus-and-the-original-date-of-christmas/.

[26] *On the Trinity* IV, 5; *NPNF*1, III, 74.

sustained—let alone his suggestion of Jesus' birth in AD 9. Consequently, it should not be too disappointing to discover that the only form of an integral age for Jesus' life that is possible to maintain, based on all the evidence from scriptural, historico-political, and calendrical sources, is that his incarnation and crucifixion both occurred on the feast of Passover—separated by a likely span of thirty-five years (3 BC—AD 33). Any suggestion of a divinely-designed integral age of Jesus beyond this appears to be an exercise in historical romanticism requiring questionable assertions about particular dates and events. Furthermore, there is quite enough evidence of divine design in the timing and execution of the events of Jesus' birth and Passion without such additional calendrical correspondence.

WHY THE EASTER CYCLE?

It has already been noted that one of the early sources of information for Christian understanding of the birth date and crucifixion date of Jesus is the set of inscriptions found on the sarcophagus of Hippolytus. The left side of the sarcophagus displays what is in fact the earliest known Paschal, or Easter cycle inscribed in stone. Being a bishop, one of Hippolytus' responsibilities was the proper scheduling of Easter in his region. Although the cycle was likely not original with Hippolytus himself, its mention by Eusebius a century later is nonetheless a testimony to its genuineness.[27]

Modern Christians might be tempted to ask, however, why Hippolytus or anyone else found it necessary to construct Easter cycles in the first place. Why were Annianus and other early Christians so vexed with the question of the calendar dates of Jesus' life? If celebrating the particular dates of Jesus' death and resurrection was so important, why not just consult the local

---

[27] Eusebius, *Church History*, VI, 22, in *NPNF*2, I, 269–270. It is likely that Hippolytus constructed the sixteen-year cycles by combining two of the eight-year cycles used by some Greek astronomers, according to which the full moon normally fell on the same day of the month once every eight years. As this is not quite correct astronomically (evidently unknown to Hippolytus), it provided the occasion for both Eusebius and Jerome to construct their own nineteen-year cycles. On the development of the *octaeteris*, or eight-year lunar and calendrical cycle by the Greeks and its eventual replacement with the more accurate *enneakaidekaeterida*, or nineteen-year cycle, see Hannah, 31ff.

synagogue for the dates of Passover and schedule accordingly? The first and most obvious answer to this question is that by the beginning of the second century circumstances had made it impossible for Christians in many places to take advantage of the Jewish connections they had once enjoyed.

Figure 10. Easter Cycle. Sarcophagus of Hippolytus (ca. 170–236). Courtesy of the Vatican Museums, Rome. Photo by Author.

Jesus' words of caution in Matthew 24:36 that, "concerning that day no one knows, not even the angels in heaven, nor the Son, but only the Father," have often been quoted as a safeguard against speculation concerning possible dates for his return. What modern Christians easily forget, however, is that these words appeared to be especially appropriate and even prophetic for Jesus' followers just a generation later. Many Christians witnessed a literal fulfillment of Jesus' words that "there will not be left here one stone [of the temple] upon another" (Matthew 24:2) when the Romans invaded Palestine to put down a Jewish rebellion in AD 70, and razed the city of Jerusalem in the process. Although their anticipation of Jesus' return at that time was to be disappointed, the destruction of Jerusalem proved nevertheless to be an important event for Christians and their relationships to both Judaism and the Roman government.

The loss of Jerusalem and the temple cult occasioned major shifts in Judaism over the next generation, although it appears that they recovered nicely by the early second century under the leadership of Rabbis Johanan ben Zakhai, Gamaliel II, and Akiba. Some Jews felt that the loss of Jerusalem to the Romans had been a manifestation of divine judgment for their lack of adherence to the Mosaic law, evidenced by the tolerance of Christians in their midst. So the remedy was two-fold: First of all, self-proclaimed Christians in Palestine were thence prohibited from attendance at synagogues, which thereby significantly decreased evangelistic opportunities for Christians among their fellow-Jews. Secondly, the canon of Jewish Scripture was defined so as to include only the thirty-nine texts known today, excluding the Old Testament apocrypha and all Christian writings. In addition, only the Hebrew text of Scripture was to be used, excluding the Septuagint which had been used effectively by Christian evangelists.[28]

---

[28] One well-known point of controversy between first-century Christians and Jews regarded the proper interpretation of Isaiah 7:14, which is quoted in Matthew 1:23 as, "The virgin will be with child and will give birth to a son"—interpreted by Christians as speaking of the uniqueness of Jesus as the Son of God. Matthew is quoting from the Septuagint, which uses πάρθενος (Greek *parthenos* = virgin) to translate עלמה (Hebrew *almah*), which simply means "maiden, young woman," not necessarily a sexual virgin (Hebrew בטולה, *bethulah*). In actuality, neither Hebrew term is strictly technical, possibly meaning either "virgin" or "non-virgin" (*Theological Wordbook of the Old Testament* [R. Laird Harris, Gleason L. Archer, Jr., and Bruce K. Waltke, eds. Chicago: Moody Press, 1980], I, 138; II, 672).

The specificity of the Christian claim of Jesus' uniqueness and the vigor of their evangelism, however, occasioned sufficient discomfort among Palestinian Jews that they felt the need for this defensive measure. The Septuagint, therefore, which once had been held in reverence by Jews as equally inspired and authoritative as the Hebrew text, came to be rejected and condemned. On the increasing rift between Jews and Christians in the late first century, see W. H. C. Frend, *Martyrdom and Persecution in the Early Church* (Grand Rapids: Baker, 1965), 178ff. Flavius Josephus tells the traditional story of the origin of the Septuagint in Alexandria in the

Rulings concerning the prohibition of Christian attendance at synagogues and proscription of all but the thirty-nine texts of Hebrew Scripture were rather slow to make it to the hundreds of synagogues across the Greek-speaking world of the Mediterranean, however, and even slower to be implemented. Knowledge of Hebrew was somewhat sparse outside of Palestine, meaning that in those regions the Septuagint enjoyed a longer life and Christians a longer relationship with their Jewish roots. Yet it was inevitable that the Jewish and Christian communities would slowly separate and pursue irreconcilable paths. In addition to their differences over the Messiahship of Jesus, from an early date the Christians began abrogating certain aspects of the Jewish law, as reflected in the debate recorded in Acts 15. Moreover, the debates about the extent to which Christians should keep the Jewish law were occasioned by their own practice of evangelism among Gentiles as well as Jews, which only further alienated some Jews who were concerned to maintain their ritual and moral purity over against the perceived filth of the Gentiles.

Speaking of Christian evangelism, it is clear from the book of Acts that after the council of Jerusalem missionaries such as Paul felt even more emboldened to preach the Gospel to non-Jews. While the historic pattern of proclaiming the message about Jesus first in the synagogues of each locale was maintained for awhile, an increasing percentage of the new converts came from Gentile rather than Jewish heritage. Again, the book of Acts itself shows us the trend which was beginning to take place within a generation after the lifetime of Jesus: The sermons of Paul before the philosophers in Athens (Acts 17:22ff) and before numerous Roman politicians on his way to a trial in Rome (chapters 24–26), coupled with his rejection by Jewish authorities and assassination attempts (21:27ff, 23:12ff), show the increasingly Gentile trend of Christianity and its increasing alienation from its Jewish roots by the second half of the first century.

Despite Paul's insistence that he had in good conscience fulfilled the requirements of the Mosaic law (Acts 23:1; Philippians 3:4–6), outright proclamation of the Christian Gospel to Gentile populations and decreasing levels of adherence to the particularities of the law by

---

third century BC and its high regard by the Jewish community in his *Antiquities of the Jews* XII, II (*The Works of Josephus* [Lynn, MA: Hendrickson, 1980], 245ff). On the Jewish denigration of the Septuagint in the late first century, see Ernst Wurthwein, *The Text of the Old Testament* (Grand Rapids: Eerdmans, 1979), 52–53.

Gentile converts were an offense to law-abiding Jews and especially to their leaders. In addition, the more the Christian Gospel spread among Gentiles, the more it spread beyond the geographical and social bounds of first-century Judaism. It therefore comes as no great surprise that Christians in Rome and elsewhere received little sympathy from their Jewish counterparts when Christians became objects of persecution by Nero in AD 64 and later by several of his successors. In fact, some Jews viewed the persecutions as opportunities to dissociate themselves from the heresy in their midst and thus to cleanse their own ranks.[29]

So by the end of the first century the inevitable and sometimes very painful divergence of Christians from their Jewish roots, occasioned at least in part by their own success in proclaiming the message about Jesus' death and resurrection, began to inhibit Christian adherence to the Jewish calendar of Passover. In fact, by the early second century there were Christian congregations in the cities of the western Mediterranean who were geographically too far afield from any synagogue, meaning that they no direct access to the Jewish calendar.

A second answer to the question of why Christians created a cycle of dates for the celebration of Easter involves developments in the Jewish calendar itself and the challenges facing Jews living outside of Palestine. The modern Jewish calendar used today was not developed until at least the fourth century AD, and possibly not in final form until the tenth century. While the Christians followed the Roman trend toward a strictly solar calendar that increasingly ignored lunar cycles except for special occasions such as Easter which required a lunar component, the Jewish calendar stubbornly retained its luni-solar character. The major reason for this seems to have been that the calendar served as a point of Jewish community solidarity. In a world in which Jews existed as scattered minority communities, occasionally facing persecution at the hands of the Romans and later even by Christians, their strict adherence to a lunar calendar provided

[29] Frend, 164ff. Frend also asserts (168–171) that the Christian community in Jerusalem declined in influence in the late first century not because it lost members who perished in the Jewish rebellion and the consequent Roman destruction of Jerusalem in AD 70, but because the majority of the Christians fled from Jerusalem to Pella, as related by Eusebius in *Church History* III, 5 (*NPNF*2, I, 138). In the minds of Jewish leaders, Christians could hardly claim to be the "New Israel" when they had not supported their Jewish brethren against the Romans.

one more way to distinguish themselves over against the perceived pollution of the Gentiles, as the one place on earth where the Mosaic prescription for righteousness reigned supreme.[30] This also shows one more example of the divergence of paths taken by Jews and Christians during an era which was formative for the future of both.

The final form of the Jewish calendar, however, was at least a few centuries in development. By the early centuries of the Christian era, moreover, due to the several historical diasporae of Jews at the hands of the Assyrians, the Babylonians, the Greeks, and finally the Romans, Jewish communities stretched from Mesopotamia in the East to Arabia and Egypt in the South and to at least Italy in the West. While Jews living in or near Palestine (or later near Babylon) had the luxury of waiting each year for the rabbis to announce whether there would be an additional lunar cycle before the beginning of the month of Abib (Aviv, or Nisan) and the Passover festival, such was not the case for their counterparts living hundreds of miles away. The slowness of travel in the ancient world, even around the Mediterranean, meant that such news would often not arrive in time for proper planning and execution of Passover. In this early period, therefore, Jews of the Diaspora adopted the Greek practice linked with the Metonic cycle: An extra lunar cycle was automatically inserted in years 3, 6, 8, 11, 14, 17, and 19 of every εννεακαιδεκαετεριδα, or nineteen-year cycle. Consequently, although there were inevitably some years when Jews of the Diaspora celebrated the Passover a month before or after the date officially proclaimed in Jerusalem, only rarely would they discover their error, and even then far too late—and therefore felt perfectly justified in their practice.[31]

---

[30] See the worthwhile treatment of this topic by Sacha Stern, *Calendar and Community: A History of the Jewish Calendar, Second Century BCE to Tenth Century CE* (Oxford: Oxford University Press, 2001), 211ff. Among other observations, Stern asserts that evidence of the pattern or procedure of intercalation (adding an extra lunar cycle) is sporadic and incomplete among Jewish sources of the early Christian era, and that in fact Christian sources are actually more informative, although their "reliability in reporting contemporary Jewish practice can sometimes be difficult to ascertain" (49). He also admits that the Christian development of Paschal cycles likely exerted some influence on rabbis of the era to solidify their own calendar (65f).

[31] On the Greek calendar, see Hannah, 35–41. Going back to the question of whether the Synoptic Gospels or John present the more accurate chronology of the date of Jesus' crucifixion, a kind of harmony between the differing accounts concerning the

This difference in calendrical use eventually affected the Christian communities as well, for it is clear from the previous discussion that by the late second century Christian leaders were intensely interested in the calendar dates of Jesus' life. Furthermore, it is clear that the dating of Passover was one piece of Jewish observance that many Christians desired to maintain, and for the obvious reason already stated: The central drama of the death and resurrection of Christ took place precisely at Passover so that its divinely redemptive intent would be clear. So while the Christians in or near Palestine could parallel their Jewish neighbors in watching for the annual announcement concerning the beginning of Nisan, this was impossible for the vast majority of Christians who were the beneficiaries of missionary efforts in the far-flung reaches of the Roman world. Like their Jewish counterparts, these Christians adapted easily enough to strict usage of the Metonic cycle. But the far-flung locations of Christian congregations and the increasingly Gentile face of the church meant that some Christians were not living in close proximity to any synagogue. This brought additional factors to bear on the calendrical matters facing Christians—matters which are equally a part of the answer to the question of why Christians began developing Easter cycles in the first place.

WHEN SHOULD EASTER BE CELEBRATED?

The debate over the proper date for celebrating Jesus' death and resurrection, of which this is only a very brief summary, proceeded in several stages over the first few centuries of Christian history.[32] On the one hand, the progress of this debate is a prime illustration of the growing Gentile nature of Christianity as compared with the decline

---

date of the Passover celebrated by Jesus and his disciples was proposed by Shepherd. Stating that John's Gospel reflects the Palestinian reckoning of Passover (based on lunar observation in Palestine) and therefore reports the true date, while the Synoptics (Matthew and Luke following the lead of Mark) reflect the practice of Diaspora Judaism in depending upon a fixed calendar, Shepherd contends that Jews in Palestine celebrated the Passover on Saturday in the year of Jesus' crucifixion, whereas Jews elsewhere celebrated it on Friday. See Massey H. Shepherd, Jr., "Are Both the Synoptics and John Correct About the Date of Jesus' Death?," *JBL* 80, No. 2 (June 1961): 123–132.

[32] More detailed accounts are given by Bonnie Blackburn and Leofreanc Holford-Strevens, *OCY*, 801ff; E. G. Richards, *Mapping Time: The Calendar and Its History* (Oxford: Oxford University Press, 1998), 354ff; Steel, 95ff.

of distinctly Jewish influence. More than that, however, it is illustrative of the growth of Christianity itself in the late Roman world, showing not only its changing relationship with Judaism and with the Roman political system, but also how Christians learned to use the intellectual and scientific resources at their disposal to inform theological questions.

The first stage in the Easter debate was over the day of the week upon which it should take place. Clearly, Christians living in the eastern Mediterranean were closer to the Jewish roots of the faith than those living farther away. In one well-known instance, moreover, insistence on faithfulness to the date of Passover superseded other scriptural input on the date of Jesus' resurrection. In the second century a controversy arose because of the practice of Christians in the Roman province of Asia (western Turkey, the region of the seven congregations to whom John wrote in Revelation 2–3), who ended their fast and celebrated *Pasch*, or *Pascha*, on the fourteenth moon (eve of the fourteenth day) of Nisan, since it was the biblically prescribed date for Passover (Exodus 12:1–11). Insistence on the fourteenth moon (*quartodecima* = fourteen), regardless of the day of the week, caused this debate to be known as the "Quartodeciman" controversy.[33]

It should be clarified here that "pascha" (πασχα) is not a Greek translation of "Passover" (Hebrew פסח, "pesach"[34]), despite their similarity in sound and their obvious historical and theological overlap. It derives rather from "pascho" (πάσχω), meaning "to suffer." It became the common term in both Greek and Latin for "Easter," and was later copied into other languages of the Latin family. Since Pascha, however, was a celebration of not only the suffering and death of Christ, but also of his resurrection, Christians in other regions insisted that the proper day for celebrating the resurrection was the Sunday following Passover, in agreement with

[33] It appears likely that Quartodecimanism also had numerous followers in Egypt. See S. Kent Brown, "An Easter Calendar on Limestone," *Acts of the Fifth Inter-national Congress of Coptic Studies*, 3 vols. Washington: International Association of Coptic Studies,1992.

[34] פסח, or "pesach" literally means to "pass over" or "pass by" as a means of withholding judgment. See *TWOT* II, 728–729.

the Gospels which state that Jesus rose from the dead on the "first day of the week."[35]

Perhaps the most charitable way of viewing this debate is to understand that for Christians in Asia the central focus in the divine drama was the suffering and death of Christ, whereas Christians elsewhere chose to focus on his resurrection from the dead. Not everyone at the time, however, was so charitable. Eusebius of Caesarea gave the essential details in his *Church History*: Sometime after AD 150 bishop Polycarp of Smyrna went to Rome to discuss the matter with Roman bishop Anicetus, but it seems that neither persuaded the other to adopt his practice. Later in the same century when Victor presided at Rome, he threatened to excommunicate Polycrates and the other bishops of Asia for their aberrant practice after receiving a remonstrance from them which claimed that their practice had originated with the apostles Philip, who was buried at Hierapolis, and John, who was buried at Ephesus. Fortunately, Victor was restrained by the irenic words of Irenaeus.[36]

A relatively final decision on the matter was made at the Council of Arles in AD 314 and reaffirmed at Nicaea in 325[37], asserting that the resurrection of Christ should be celebrated only on a Sunday, although some Christians in Asia continued their quartodeciman

[35] A related matter of importance which nevertheless need not be a distraction to this discussion is that of the origin of Christian worship on Sunday. While there is not enough biblical evidence to support the contention of some that it stems immediately from the Resurrection appearances of Jesus, there is evidence that Christians did at least sometimes meet on Sundays, perhaps as an addendum to worship at synagogues for the purpose of specifically Christian teaching and fellowship (Acts 20:7). More decisive is evidence from the second century which unmistakably identifies Sunday as the universally observed day of Christian gatherings for worship, with no implication that there had been any controversy over the matter previously. See Richard J. Bauckham, "The Lord's Day," *From Sabbath to Lord's Day: A Biblical, Historical, and Theological Investigation* (Donald A. Carson, ed. Grand Rapids: Zondervan, 1982), 236ff. A contra argument—thoroughly rebutted by Bauckham—was attempted by Samuele Bacchiocchi in *Anti-Judaism and the Origin of Sunday* (Rome: Pontifical Gregorian University Press, 1975).

[36] *Church History* V, 23–25, in *NPNF*2, I, 241ff.

[37] *NPNF*2, XIV, 54. The modern reader will no doubt note what appears to be an anti-Jewish tone in the words of Constantine—words which have sometimes come back to haunt Christians more recently. Without trying to defend Constantine, we should note that this is not only a reflection of Christianity's rise to power in the late Roman world, but also of the bitter memory of many Christians who were ousted from Jewish communities.

practice for another century. In retrospect, it might be said that this was another example of the Gentile trend of Christianity slowly displacing and replacing its original Jewish character and practices. At the same time, however, it is difficult to disagree completely with Hippolytus, who perceived in the Quartodeciman practice an attempt to impose Jewish law upon Gentile Christians, among whom the true Passover "is discerned by faith, and not now observed in letter (merely)."[38]

But if Easter is to be celebrated on the Sunday after Passover, and Passover is the fourteenth day of Nisan, and Nisan begins with the new moon nearest the vernal equinox, then exactly what Sunday is it? And what authority was to determine the proper Sunday and publish it well in advance for the benefit of those Christians living far from Palestine and the annual announcement of the beginning of Nisan and the Passover? Although these questions are not answered by the documents from Nicaea, it is obvious from the "Festal Letters" written by Athanasius annually from AD 328–373 that the task of determining the proper date was given to astronomers in his diocese at Alexandria. Perhaps this was meant as a kind of Christian equivalent to the Jewish dependence on annual announcements of the Sanhedrin, but the problem remained the same: Unless calculations were made and letters were sent out well in advance, those living at a distance would be uninformed and therefore could only guess at the proper date for Easter. A second and related difficulty for Christians living at a distance from Alexandria was the fact Athanasius' bishopric was interrupted by no less than five periods of exile from his post, due at least partially to the continuing popularity of Arianism despite its condemnation at Nicaea.[39]

---

[38] Hippolytus, *Adversus Haereses*, VIII, 11, in *ANF* V, 123. See also Shepherd, *The Paschal Liturgy and the Apocalypse*, (Richmond, VA: John Knox Press, 1960), 41ff, who asserts that the Quartodecimans likely reflect the practice of the first generation of Christians. For more on the Quartodecimans, including some of their own statements, see *Easter in the Early Church: An Anthology of Jewish and Early Christian Texts* (Raniero Cantalamessa, ed. Collegeville, MN: Liturgical Press, 1993), 33–37, 46–50.

[39] For the annual "Festal Letters" of Athanasius, see *NPNF*2, IV, 506ff. Athanasius was one of the leading orthodox voices at Nicaea, and served as bishop of Alexandria AD 328–373. His dates for Easter also make it clear that 21 March was the assumed date of the vernal equinox in Alexandria. His orthodoxy and authority in determining the date of Easter were affirmed at the Council of Sardica (AD 343), which was

It is therefore no surprise that a variety of forms of the *computus*, or *cyclus paschalis* (Easter cycle) were proposed and debated, and that they retained popularity for generations.[40] At the long end of these was the aforementioned 532-year cycle of Annianus, although most preferred to use the shorter and more manageable cycles of 84 or 112 years designed at Rome, or the 95-year cycle designed at Alexandria. Certainly one of the earliest paschal calendars at Alexandria developed around Metonic nineteen-year cycles was that of Anatolius (d. 282)—a native of Alexandria who later became bishop of Laodicea.[41]

In the interest of cooperation against the Arian heresy, ecclesiastical authorities at Rome and Alexandria managed in some years to compromise with one another on the proper date for Easter Sunday, but unfortunately each had their problematic limitations in computation: Astronomers at Alexandria dated the vernal equinox to 21 March, whereas those at Rome sometimes marked it as early as 18 March. The Alexandrians objected that anything before 21 March was too early because an Easter before 21 March would actually be in the twelfth month and therefore even in the wrong year (since the spring equinox was viewed as the beginning of the year). But the Romans objected that the Alexandrian computation allowed for Easter to be sometimes celebrated after 21 April, which made it in the second month of the year and therefore unacceptably late.

In addition, the Alexandrians stipulated that Easter had to be celebrated somewhere between *lunae* 15–21 (the fifteenth to twenty-first days of the lunar cycle) so that it would occur within a week after Passover (14 Nisan), since Scripture is clear that Jesus rose from the dead on the Sunday following Passover. The Romans, meanwhile, insisted on lunae 16–22, contending that since Jesus died on a Friday and rose on a Sunday he could not have risen from the dead as early as 15 Nisan. This meant, however, that if 15 Nisan was a Sunday

---

intended to be ecumenical but was boycotted by some Eastern bishops (*NPNF*2, XIV, 433–434).

[40] The most comprehensive study of early Christian Paschal cycles is Mosshammer's *The Easter Computus,* 109ff.

[41] Eusebius, *Ecclesiastical History* VII, 32 (*NPNF*2, I, 319). He also mentions that the Jews were careful to celebrate Passover only after the vernal equinox. It is believed that Anatolius introduced his paschal calendar about AD 258. See Faith Wallis, *Bede: The Reckoning of Time* (Liverpool: Liverpool University Press, 1999), xlvii.

(unacceptably early for the Romans) Easter could not be celebrated until 22 Nisan, which was more than a week after Passover and therefore unacceptable to the Alexandrians. Needless to say, in spite of the greater theological threat from Arianism their cooperation was sporadic and short-lived.

A merger of sorts between the two sides appears to have been attempted by Victorius of Aquitaine with his *Cursus Paschalis* in AD 457. While using the Alexandrian (Metonic) methods of tracking lunar cycles, he nevertheless used the Roman observations of the age of the moon and retained the Roman lunar limits of dates for Easter. This apparently wide-minded and irenic approach, combined with his construction of a 532-year master cycle of dates for Easter, must have appeared initially impressive to ecclesiastical authorities, covering the period AD 28–559 (*Anno Passionis* 1–532, dated from Victorius' supposed date of Jesus' crucifixion and resurrection). Nevertheless, its use of differing systems of information, combined with other small historical inaccuracies, inevitably produced considerable confusion from both Roman and Alexandrian perspectives. Moreover, although in those years in which there was some question as to the correct date of Easter he included both early (Roman) and late ("Greek") lunar figures and left the final decision to the Pope, in at least one year (AD 482) he violated the lunar limits of both Rome and Alexandria and incorrectly had the "Greeks" observing Easter on a Saturday.[42] It was at least partially because of such inaccuracies and associated difficulties with Victorius' cycle, therefore, that Dionysius Exiguus was enlisted by the Papacy to create a new Paschal cycle less than a century later.[43]

---

[42] The Paschal cycle of Victorius is available at *Monumenta Germaniae Historica*, http://www.dmgh.de/de/fs2/object/display/bsb00000798_00698.html?sortIndex=010:010:0009:010:00:00. Victorius' calculations for AD 482 are found on page 727. For the reader's convenience, a comparison of the lunar measurements between Alexandria and Rome, along with their impact on dates for Easter, is given in Appendix D. Although it covers only the years included in the Paschal cycle of Dionysius (AD 532–626), the difference in lunar measurements is immediately apparent and their impact on the dates of Easter unavoidable. See also *OCY*, 793f, 807ff.

[43] See also Bruno Krusch, *Abhandlungen der Preußischen Akademie der Wissenschaften. Philosophisch-historische Klasse, Nr. 8: Studien zur christlich-mittelalterlichen Chronologie* (Berlin: Walter de Gruyter, 1938), 59; Arno Borst, Die karolingische Kalenderreform (Hannover: Hahnsche, 1998), 177.

The work of Dionysius and its impact is the focus of the next chapter. Before considering the work of Dionysius and his possible reactions to Victorius, however, it should stand out to us as truly remarkable that Victorius even dared to use astronomical measurements taken from Rome in the first place. Although not stated specifically in the documents from Nicaea, it is clear from several sources—from the early Paschal Cycle of Anatolius more than a century before Nicaea, to the Paschal Letters of Athanasius in the generation immediately following Nicaea, to the Paschal cycle of Dionysius two centuries later—that the intended authority for establishing the dates for Easter was to be the lunar observations taken from Alexandria. One of those pro-Alexandrian voices in the years just preceding the work of Victorius was Paschasinus, bishop of Lilybaeum (modern Marsala, Sicily), who in AD 444 responded to an inquiry from Pope Leo the Great concerning the proper dates for Easter by recommending that he use the Alexandrian calculations.[44] That Paschasinus should ignore his nearer proximity to Rome and risk offending his spiritual superior for the sake of a scientifically accurate answer concerning the most important of Christian festivals is perhaps remarkable enough. But that Victorius, less than twenty years later, should ignore the several indications of Alexandrian accuracy, apparently for the sake of pleasing his superiors at Rome—with the unintended result of giving them a patently false sense of security in their own calculations—is nothing less than astonishing.

---

[44] *NPNF*2, XII, 2. A possible reason for Paschasinus' recommendation of the Alexandrian dates is the legend of the baptismal font at Meltinas in Sicily, which reportedly filled spontaneously with water every year at the vigil of Easter just in time for the baptism of new converts. Curiously, it is reported that the font was in the habit of filling only in accordance with the Alexandrian dates of Easter, but not the Roman dates (Mosshammer, 63). For more examples of Christian leaders in the generations after the Council of Nicaea who voiced preference of Alexandrian dates for calculating the vernal equinox and Easter celebration, see Hefele's brief essay in *NPNF*2, XIV, 55f.

5

Dionysius Exiguus, the Paschal Cycle, and the Advent of the Christian Era

OR,

Little Dennis and His Very Big Job

The early medieval era has been often described as the "Dark Age," and with good reason. Stretching roughly from the fall of Italy and the city of Rome to Odoacer and the Ostrogoths in AD 476 up to at least the coronation of Charlemagne as the first Holy Roman Emperor on Christmas Day in AD 800, it was undoubtedly the *darkest* cultural period the west has known in its recorded history. Imperfect as the Romans were, they had nevertheless succeeded in bringing relative peace and security throughout most of the Mediterranean region most of the time for the previous several centuries. Certainly not every person in the widely varied ethnic groups of the Mediterranean liked the Romans, and there were occasional revolts—at least a few of them rather serious. For the majority of people, however, who were not inclined to take up arms but rather to live peacefully and put up with the Roman system, there were the advantages of economic prosperity brought by the free flow of goods, people, and services throughout the region.

But the peace and prosperity were gone by the late fifth century. The Roman legions had slowly retreated eastward to defend and build up their new capital city at Constantinople. In their wake, the once peaceful Roman dominions of western Europe and northwest Africa were ravaged again and again by the semi-nomadic Germanic tribes from northern Europe, and still others from central Asia. Among the cultural victims in this onslaught were nearly all manner of scholarship in the arts and sciences, which are dependent upon a relatively stable political environment in order to flourish. The few intellectual pursuits which did survive were those directly connected to the Christian church—the one Roman social institution which remained in the west, and which was ultimately successful in slowly converting the "barbarian"[1] tribes not only to the Christian message but also to numerous other aspects of the old Roman cultural system. Ironically, it could be said that the Christian church survived and even thrived among the barbarian tribes of the west because—in spite of the official status it gained in the Roman world within a couple generations after the conversion of Constantine (AD 312)—the ultimate loyalty of Christians was to the Creator and Redeemer of all humanity rather than to any political system.

An unfortunate but popular myth about the Middle Age(s) which has been fostered by some movies and books is that the entire period—from the fall of Rome in the late fifth century to the Renaissance and the beginnings of the European age of exploration and the Protestant Reformation nearly a thousand years later—was one of unmitigated cultural darkness. A common assertion which dates back at least as far as Edward Gibbon's *The Decline and Fall of the Roman Empire* (1776) is that, with the conversion of Constantine, Christianity became a mere servant to the state, or vice-versa, in their mutual quest for power over the lives and souls of humanity. Furthermore, with its promise of eternal blessedness in the kingdom of God, Christianity purportedly weakened the resolve of Romans to deal with the challenges of their age, with the result that the Roman world of the western Mediterranean fell to the barbarian tribes. A similar assertion is that by the early fourth century AD Christianity had triumphed in Roman society at the expense of the formerly influential Greco-Roman philosophies. The medieval era is therefore to be viewed as nothing less than a long "dark age" in which the learning of the ancients was deliberately discarded or obscured, and

[1] "Barbari" and "pagani" were basically synonymous terms, both used in a pejorative sense by the Romans to mean "uncivilized," as a descriptor of those tribes living outside or at the fringes of Roman civilization.

any attempts at furthering learning and bettering the human condition were suppressed by the politically power-hungry Roman Catholic Church.[2]

Fortunately, while containing some small elements of truth, this rather anti-Christian reasoning fails on two major counts. First of all, the assertion that Christians as far back as the apostle Paul attempted to squelch Greco-Roman philosophy betrays an ignorance of the text of the New Testament itself, in which Paul's dependence on Greco-Roman philosophical categories to explain the Christian message to the Athenians (Acts 17:16–34) and the Colossians (Colossians 2:1–12) is all too obvious. The same can be said for the anti-Gnostic comments found in the Gospel of John (John 1:1–5) and the letters of John (especially 1 John 1). Likewise, it ignores the open embrace of philosophy by later Christian leaders such as Augustine, whose concept of the two cities is modeled after Plato's *The Republic*.[3] Secondly, this assertion does a disservice to the labors of numerous individuals, such as Dionysius Exiguus, who is the central focus of this chapter, and numerous others who preserved, copied, and translated ancient texts—far beyond just that of the Bible—so that they might be transmitted to succeeding generations in spite of the unstable political situation of the fifth through eighth centuries. Without the faithful contributions of individuals such as Dionysius, it is likely that the west would have been an even darker place and for a longer period of time.[4]

A very different argument has been constructed by other authors who view the medieval institutions of church and state as having incalculable value as foundation stones or predecessors of the present. These authors contend that Christianity laid the essential philosophical groundwork for technical, political, and social advances during the Middle Ages because of its positive view of God, providence, and human ability to understand the divine plan. They further state that Christianity has continued its geographical

[2] Charles Freeman, *The Closing of the Western Mind: The Rise of Faith and the Fall of Reason* (New York: Vintage, 2005). Although only a sub-theme, this titillating but historically inaccurate view of Christian leaders as power-hungry zealots was popularized by Dan Brown in his best-selling novel, *The DaVinci Code*.

[3] Augustine, "City of God," *A Select Library of the Nicene and Post-Nicene Fathers of the Christian Church* (Philip Schaff, ed. Edinburgh: T & T Clark, 1988 [1886]).

[4] David Ewing Duncan, *The Calendar* (London: Fourth Estate, 1998), 79ff.

expansion into the present in regions such as Africa, China, and Latin America (usually without the sword!) because of its rational theology, insistence on moral equality, and inherent faith in the divinely-determined future.[5] Similarly, it has been asserted that it is in medieval Catholic theology and philosophy that we find the roots of modern science, free-market economics, the western legal system, and the modern university. Furthermore, the rule of law, the assertion of individual human rights and freedoms, philosophy grounded in reason, and charity as a virtue were all shaped and nurtured in the medieval Catholic tradition.[6]

Enough of the debates about the place of Christianity in the late ancient and medieval worlds. Whatever level to which our modern society is indebted to the medieval era and its institutions or, conversely, whatever the level of retrenchment and cultural darkness attributed to it, what can certainly be said is that although *darkness* may be an appropriate descriptor for the early medieval period, there were definitely some bright lights shining in the midst of the relative darkness. This chapter focuses on one of those bright lights, an individual whose life might have been easily consigned to the dustbin of human history had he not created a body of work which somehow caught the attention of those living after him. We know really nothing of his life outside his extant works, and have no evidence that he received any recognition for his work during his own lifetime. Yet in the long run he became one of the most influential Christian scholars of the era.

His name, as stated already, is Dionysius Exiguus (ca. AD 485–545), or "Dennis the Small One" (likely a title of humility), a monk from Scythia (modern Romania and Ukraine) who apparently lived his adult life in and around Rome. We are told by his contemporary Cassiodorus that he was a faithful servant of the Lord and the church by virtue of his translation of numerous books from Greek into Latin.[7]

---

[5] Rodney Stark, *The Victory of Reason: How Christianity Led to Freedom, Capitalism, and Western Success* (New York: Random House, 2006).

[6] Thomas E. Woods, Jr., *How the Catholic Church Built Western Civilization* (Washington: Regnery, 2005).

[7] *Cassiodori Senatoris Institutiones* (R. A. B. Mynors, ed. Oxford: Clarendon Press, 1937), 62. Cassiodorus also provided us with the earliest known use of the Dionysian Paschal cycle in a short set of instructions on time measurement which he wrote in AD 562 for monks at his school in southern Italy. See Georges Declercq, *Anno Domini: The Origins of the Christian Era* (Turnhout: Brepols, 2000), 150–151.

Justification for Cassiodorus' praise is seen in Dionysius' compilation of the first comprehensive edition of canon law, although he is perhaps known best for the subject at hand—his chronological works. In the nearly 1,500 years since its initial construction, the Dionysian chronology has been occasionally the focus of scholarly investigation, and was the focus of more popular attention during the twentieth century and especially near the turn of the twenty-first century and third Christian millennium.

From his prefatory letter to an otherwise unknown papal representative by the name of Petronius and his *Argumenta Paschalia*, it is clear that Dionysius began his chronological work about AD 525. His task—apparently assigned to him by the papal office—was the construction of the next ninety-five-year cycle of dates for the celebration of Easter. He began his task rather simply by extending the nearly-expired Easter cycle attributed to Cyril of Alexandria, which had covered the years *Anno Diocletiani* 153–247—years since the beginning of the reign of Roman Emperor Diocletian.[8] Dionysius appended the final Metonic cycle of Cyril's Easter table (Anno Diocletiani 229–247), and then began his own with Anno Diocletiani 248. But this is also the point at which he also began the monumental change in chronological orientation to *Anni Domini Nostri Jesu Christi*, or "year of our Lord Jesus Christ"—which most of us know by the shortened form *Anno Domini*, or "in the year of our Lord." As Dionysius explained in the letter to Petronius, there was no good reason for Christians to continue dating themselves by the reign of an emperor who had in fact been not only a pagan but their persecutor while, on the other hand, there was every reason to date themselves by the incarnation of God into human flesh.[9] So the first document and mathematical calculation based on the historical occurrence of the life of Christ was, very appropriately, a table of dates for the celebration of his resurrection from the dead.

---

[8] Diocletian's reign began in AD 284, according to our counting using the system created by Dionysius. Cyril's Easter table covered *Anno Diocletiani* 153–247, or *Anno Domini* 437–531.

[9] Dionysii Exigui, "Epistolae Duae de Ratione Paschae," *PL* (Montrouge: Parisiorum, 1848), LXVII, 19ff.

Whether Dionysius intended any use of this designation outside the dating of Easter is entirely debatable.[10]

"LITTLE DENNIS" KNEW EXACTLY WHAT HE WAS DOING (I THINK)

Questions related to Dionysius' Paschal cycle and his chronological calculations, as well as his underlying motives, have been the matter of debate among scholars for the last century or more. Some have focused on his chronological calculations and their underlying astronomical observations, while others have focused rather on his theological perspectives. Certainly not everyone has shared an attitude of reverence toward Dionysius and his work, as one recent author described his chronology as a rather laughable matter guided more by pagan myths than Christian theology. Another author complained that Dionysius' chronology has afforded Christianity little more than confusion and dissension, but added that his most egregious decision was to not include a year *zero* between BC and AD (surely a misapprehension of Dionysius' system, in the opinion of this writer). And even the insightful scholar Joseph Scaliger thought it a scandal when writing in the late sixteenth century that Dionysius' miscalculation of Christ's birth afforded Jews and unbelievers a chance to mock Christianity.[11]

---

[10] Appendix C gives an English translation and explanation of the details of Dionysius' Easter cycle (AD 513–626). Appendix E is a reconstruction of the Julian and Hebrew calendars of AD 513–569—the first three Metonic cycles of the period covered by Dionysius' cycle.

[11] James A. Veitch, "Editorial: Dionysius Exiguus," http://www.westarinstitute.org/Periodicals/4R_Articles/dionysius.html; Stephen Jay Gould, "Dousing Diminutive Dennis' Debate," *Natural History* 103, No. 4 (April 1994): 4–8; Athony T. Grafton, *Joseph Scaliger: A Study in the History of Classical Scholarship* (Oxford: Clarendon Press, 1993), II, 312. Michael Siemon asserted that Dionysius intended a kind of zero-calculation by designating 25 December 1 BC as the date of Christ's birth (http://www.cs.rutgers.edu/pub/soc.religion.christian/faq/easter-date). Veitch and Gould also mistakenly accuse Dionysius of a "mistake" in not including the year "0" in his chronology, even while acknowledging that such a number was not yet commonly used in the west. This is hardly a "mistake" or error for Dionysius, but rather a failure on the part of some modern authors to correctly apprehend his system and grasp the fact that it makes perfect sense that the year "2000," or "Twenty-Hundred," if you will, was by design the last year of the twentieth century and the second Christian millennium. After all, how could we have called it the "twentieth" century if the prefix "20__" never showed up in it?

What is often missing, however, in many of the critiques of Dionysius and the general interest articles on chronology, is an appreciation of the intense piety which guided his writing and calculations. As if his very subject matter—the celebration of Christ's resurrection from the dead—is not evidence enough, several other indications of his faith are given in the letter to Petronius. In addition to his aforementioned explanation of the shift in chronological orientation from *Anno Diocletiani* to *Anno Domini*, Dionysius claims to be in that line of computists whose calculations have been guided by illumination from the Holy Spirit. In a show of faithful churchmanship, he affirms that he has followed the practices of Alexandrian scholars, whose observations of the Paschal full moon had been the recognized standard for determining the date of Easter since at least the Council of Nicaea two centuries earlier. And perhaps there is a hint of his hope that his chronology will allow Christians to put the stamp of their faith on the whole of human society when he asserts that this will "cause the redemption of humanity through the passion of our Redeemer to shine more brightly."[12]

DIONYSIUS AND THE 532-YEAR CYCLE

What is especially impressive in the view of this writer is the mastery of Dionysius in bringing together his Christian faith with his skill as a computist—a mastery which is perhaps all the more evident when one considers the several matters under debate. One of those debates, of course, is whether Dionysius was aware of his apparent error in calculating the date of Jesus' birth as 25 December, 1 BC. Numerous scholars going at least as far back as Abbo of Fleury and Heriger of Lobbes at the end of the first Christian millennium—and possibly even as far back as the Venerable Bede in the early eighth century—have noted this apparent miscalculation, since it postdates by a year or more all of the accepted possibilities of the date of the death of Herod the Great in Judea, and therefore also the accepted

---

In fact, Dionysius never named the years preceding Christ's birth as "BC." Furthermore, although Dionysius did not insert a year "zero" before AD 1, Oppert notes that he is ostensibly the first Latin writer to use "nulla" (nothing) as a mathematical calculation. See Gustav Oppert, "Über die Entstehung der Ära Dionysia und den Ursprung der Null," *Verhandlung der Berliner Anthropologischen Gesellschaft* (Berlin: A. Asher, 1900), 122ff.

[12] *PL* LXVII, 19–20. Author's translation.

possibilities for the birth of Jesus.[13] It has also been said that Dionysius had no hard evidence for his calculation that the birth of Christ occurred 532 years before the first year of the Easter cycle he was about to create, and that he was focused rather on a best set of dates for the Easter cycle to fit his predetermined formula.[14]

While these are certainly legitimate questions, they can serve as unnecessary distractions from the main thrust of Dionysius' work, which was calculating the proper date for celebrating Christ's resurrection from the dead. On the question of his possession of hard evidence, a possible clue is the fact that one column of his Easter cycle and additional space in the supporting *Argumenta Paschalia*[15] are dedicated to the *Indiction*—the fifteen-year cycle of taxation and secular office holders in Rome. While the Indiction[16] had nothing directly to do with the calculation of dates for Easter, it does tell us something about Dionysius' sources of information. It is hence entirely plausible that he had access to the consular histories of Rome going all the way back to the reign of Augustus Caesar and the birth

---

[13] Declercq, 189–190; Bede, *De Temporae Rationem*, 47.

[14] Sepp Rothwangl, "Consideration of the Origin of the Yearly Count in the Julian and Gregorian Calendars" (http://cura.free.fr/xx/17sepp1.html) makes the very interesting but highly speculative assertion that Dionysius' calculations of the birth date of Christ originated with his desire to relieve his contemporaries of the fear of the impending end of the world according to the ancient system of Anno Mundi (AM), as described in chapter one. According to Rothwangl, Dionysius calculated AD 1 by "using the conjunction of all the planets in the year 531 CE and the commensurate planetary periods to calculate the date of a future occurrence of the same conjunction. After finding the future alignment of all planets (which occurred in May 2000), the starting point of the new yearly counting was established exactly 1999 years before it, by determining the length of one [astrological] age, using the constant of precession (66.6y/1°)." He admits, however, that nowhere in Dionysius' chronological writings does he ever mention the matter of Anno Mundi or give hint of such observations.

[15] It should be added that only the first nine of the sixteen *Argumenta* found in the standard edition of Dionysius' works are of his authorship, as only the first nine appear in a manual for paschal calculation written by his friend Cassiodorus in 562. *Argumenta* 11–12 imply that they were written about AD 675. Furthermore, several factual errors and contradictions in logic are found in *Argumenta* 14–15.

[16] While the particular form of the Indiction used during Dionysius' lifetime was instituted by Diocletian and revised by Constantine, it is clear from the *Chronicon Paschale*, a seventh-century Byzantine history, that it was based on the earlier form of Indiction used since 42 BC. See *Chronicon Paschale* (Michael Whitby and Mary Whitby, eds. Liverpool: Liverpool University Press, 1989), 1; Blackburn and Holford-Strevens, 769f.

of Jesus, in spite of the downfall of Roman rule in the western Mediterranean and the ups and downs of their Germanic successors in the half-century preceding his own work.

On the matter of the 532-year cycle for calculating Easter, some have concluded that Dionysius was simply unaware of its utility because he never mentioned it specifically. This omission seems all the more incredible to modern eyes, given the fact that Dionysius began his Paschal cycle in AD 532, which was the first year of the second master calendar cycle since the birth of Jesus. While we may never know the answer conclusively, what is obvious in the *Argumenta* is that Dionysius clearly did grasp the utility of four, seven, and nineteen—the denominators of 532. In addition, a reasonable hypothesis in this regard is that Dionysius wished to give no credit to the aforementioned Victorius of Aquitaine for his work, since the dates in Victorius' cycle were based upon astronomical observations taken from Rome rather than the more highly regarded observatory at Alexandria.

In the previous chapter the rivalry between Alexandria and Rome over lunar measurements determining the dates for Easter was reviewed, noting that most authorities from the Council of Nicaea onward—except perhaps those very near Rome—preferred to utilize the measurements from Alexandria. Also observed was the apparent attempt at a merger between the two sides by Victorius of Aquitaine with his *Cursus Paschalis* in AD 457, which unfortunately confused matters with historical inaccuracies and use of differing systems of information, resulting in conflicting and sometimes incorrect dates for Easter. As already noted, it should strike us as astonishing that Victorius even dared to use lunar measurements from Rome, given the multiple voices which had already pointed toward Alexandria. While his use of Roman data was undoubtedly pleasing to his spiritual superiors at Rome, the unintended effect was very possibly that of giving them a false sense of security.

It is therefore quite likely that it was at least partially because of the inaccuracies and difficulties associated with Victorius' cycle that Dionysius was enlisted by the Papacy to create a new Paschal cycle less than seventy years later. Even if that is not the case, it is certainly clear that Dionysius had ample reason to reject the work of Victorius as one of the "aforementioned errors" rebuked in his letter to Petronius. Why? A comparison of the Paschal dates of Victorius with those of Dionysius for the period covered by Dionysius' cycle (AD 532–626) reveals an interesting pattern: In every Metonic cycle the first six years are in complete agreement on the date of Easter and the age of the moon on that date. Beginning with the seventh year of each cycle, however, Victorius' age of the moon (taken from Roman measurements) becomes one day ahead of those on Dionysius' cycle (taken from Alexandria), with the result that differing dates for Easter occur about once in every Metonic cycle. In other words, the less accurate measurements taken from Rome, combined with differing limitations for the age of the moon at Easter (lunae 15–21 was required by the astronomers at Alexandria, lunae 16–22 by those at Rome), caused an irreconcilable disagreement of dates on a regular basis.[17]

Since it is abundantly clear from his Easter cycle that Dionysius understood the computistical functions and the vagaries of both lunar and solar cycles, it is therefore unthinkable that he did not grasp the utility of the 532-year master calendar cycle as well. Hence it is the opinion of this writer that Dionysius most certainly did understand the master calendar cycle, which allowed one to synchronize lunar dates, solar dates, and days of the week.[18] Why else would he begin his cycle with AD 532, if not to highlight the significance of the number? Then why did he not mention it specifically in his *Argumenta*, or in the letter to Petronius? The most likely explanation is that he found himself walking a political tightrope in avoiding offense to his ecclesiastical superiors in Rome—some of whom may have been ardent admirers of Victorius. After all, as seen in the closing section of this chapter, the work of Victorius retained popularity in some regions for at least another century. Since Dionysius had already admitted to using the lunar calculations from Alexandria rather than Rome, he likely did not wish to further highlight his disagreement with Victorius by specific discussion of the 532-year cycle for which Victorius was well known. Consequently, he described competing

---

[17] Appendix D details the differences in lunar figures between Victorius and Dionysius. The *Cursus Paschalis* of Victorius is available through *Monumenta Germaniae Historica* at http://www.dmgh.de/de/fs2/object/display/bsb00000798_00698.html?sortIndex=010:010:0009:010:00:00. Dionysius' *Liber de Paschate* is found in *Patrologia Latina* LXVII and at http://hbar.phys.msu.ru/gorm/chrono/paschata.htm.

[18] This is in agreement with Oppert, 116–118.

paschal cycles merely as "errors," and did not dare to mention Victorius by name.

Not all scholars have agreed with this reasoning, of course. For instance, both Declercq and Mosshammer have asserted that Dionysius deliberately employed faulty mathematics in order to create the date for the beginning of his cycle. Specifically, citing a prologue to an Easter cycle created by Felix Gillitanus for AD 627–721 (the next ninety-five-year cycle following that of Dionysius) in which Felix referenced Dionysius as the source of his computations, they contend that Dionysius took advantage of the lack of computistical skill among his western counterparts by asserting that there were only 284 years between the birth of Jesus and the beginning of the reign of Diocletian when he should have known very well from Jerome's Latin translation of Eusebius' *Chronicle* that the correct number of years was 287. His deliberate tampering with the data then allowed him to date the beginning of his own cycle at AD 532—one complete great Paschal cycle past the birth date of Jesus.[19]

While this theory is certainly attractive, it fails on two counts. First of all, it is self-contradictory, claiming on the one hand that Dionysius failed to comprehend the utility of the 532-year master Paschal cycle, while on the other hand claiming that he was attracted irresistibly to the number 532 because for his task at hand 532 was obviously the ideal number, and therefore the ideal number for the first year of his new Easter table.[20] But why was 532 the ideal number for Dionysius, except that he understood it as the multiple of the numbers nineteen, seven, and four? And why was he attracted to it as the multiple, except that he understood the necessity of synchronizing nineteen (the number of the Metonic lunar cycle), seven (the number of the days of the week), and four (the number of the leap year cycle)? And why did he understand the necessity of synchronizing these three numbers, except that he understood the great Paschal cycle?

Secondly, this assertion makes Dionysius into a clever deceiver and manipulator of the numbers, which hardly squares with Dionysius' own claim that his work was accompanied by the illumination of the Holy Spirit. So if we wish to take Dionysius at all seriously as a man of faith, it is more consistent to follow the

---

[19] Declercq, 141–143; Mosshammer, 349–350.

[20] Declercq, 140, 141.

suggestion of Bond that Dionysius misinterpreted Clement's statement of Jesus' birth in the twenty-eighth year of Augustus' reign as measured from his recognition by the Roman Senate in 27 BC (727 AUC) rather than his victory over Antony and Cleopatra in 31 BC (723 AUC).[21] Consequently, it is a reasonable conclusion that Dionysius indeed understood the 532-year cycle, regardless of whether he mentioned it specifically.

DIONYSIUS AND THE ESSENTIAL DATES OF JESUS' LIFE

Going back to the matter of Dionysius' alleged miscalculation of the year of Jesus' birth, it must be remembered again that the prime basis of his calculations from the beginning of his task in AD 525 was not the calendar year of Jesus' birth, but rather the calendar year of his death and resurrection. This corrected point of view shows the genius of Dionysius and allows one to view the purposes of the cycle from the faith perspective of a sixth-century Christian. The prime focus should not be on the beginning date of Dionysius' cycle—532 years after his ostensibly chosen date for the birth of Christ—but rather on the date of AD 563, which would appear to be 532 years after his chosen date for the death and resurrection of Christ in AD 31. In Dionysius' Easter cycle, the date for the festival of the resurrection in AD 563 is the exact date that should be expected once the faith perspective of Christians in the late ancient world is understood—25 March.[22] As noted previously, this was the ancient date of the vernal equinox and the assumed date of the creation of the world, the incarnation of the Son of God, and finally—as seen here again—the return to life of the Son of God who offers new life to a sinful humanity through his redemptive sacrifice.

A worthwhile question, presumably, is why Dionysius' calculations were not based on the fact of Easter falling on 25 March in AD 574 or 585 rather than AD 563. This was also the date for

---

[21] Clement, *Stromata,* I, 21; *ANF* II, 333. Mosshammer contends that Dionysius could not have made such a mistake (342–343). Another possibility, of course, is that Dionysius adopted the date from the implications of Africanus, which is Mosshammer's ultimate conclusion concerning the origin of Dionysius' cycle.

[22] It should be noted that a decisive factor which likely gave Dionysius confidence in his calculation of the Anno Domini chronological scheme is the relative scarcity of the occurrences of 25 March among the possible dates for Easter. Since it is rather early in the range of possible dates, its rate of occurrence is just over 2%. See http://www.smart.net/~mmontes/freq3.html#LBY.

Easter Sunday in AD 479 and 490, during the previous Easter cycle constructed by Cyril. But the use of any of those dates as a 532-year baseline for calculating the birth date (or death-and-resurrection date) of Christ would of course result in a chronology much more inaccurate than is already the case with that constructed by Dionysius, which may again indicate that he had access to other chronological sources which prevented that greater mistake.

Speaking of other sources used by Dionysius, an alternative argument has asserted that Dionysius was not the author of the "Christian Era" at all, but that its origin is to be found in the early third-century *Chronographiai* of Sextus Julius Africanus mentioned in the previous chapter. According to Mosshammer, piecing together the evidence from Alexandrian scholars in the generations following Africanus shows a pattern of adoption and transmission of his work by scholars such as Anatolius of Laodicea, Athanasius, Andreas, Theophilus of Alexandria, Annianus, Panodorus, and finally by Dionysius Exiguus.[23] Dionysius was therefore responsible, according to this theory, not for inventing a new chronological orientation, but for the more modest achievement of establishing the Alexandrian Paschal cycle as the standard for the western church. While he is certainly correct in asserting that Dionysius and every other chronologer would take advantage of the work of predecessors such as Africanus, the evidence for such direct dependence in constructing *Anno Domini* is nevertheless scanty. Moreover, Africanus himself never specifically postulated a "Christian Era," although it is possible to infer such from his work. At the same time, however, Dionysius' matter-of-fact usage of *Anno Domini* without lengthy explanation may indeed be an indicator of widespread familiarity.

Whatever the case, the writings of Dionysius themselves give no evidence other than that he began his task by calculating the dates for Easter over the next ninety-five years (AD 532–626) by simple computation with lunar and solar cycles. That being the case, it is entirely plausible that Dionysius noticed the occurrence of 25 March

[23] Mosshammer, 385ff. Alternatively, McCarthy has concluded from the Paschal table of Eusebius (ca. AD 309) that he was in fact the first to deliberately identify the year of Jesus' birth and consequently to number the years from the event of the Incarnation. See Daniel P. McCarthy, "The emergence of Anno Domini," *Time and Eternity: The Medieval Discourse* (Gerhard Jaritz and Gerson Moreno-Riano, eds. Brepols: Turnhout, 2003), 51–53.

as the Easter date for AD 563, and subtracted 532 years to locate of the year of Christ's Passion—which then also gave him the correct number of the years in which he himself was living and working.[24] Whether he had any clue as to the number of years between Christ's lifetime and his own before he began this project is not known, but what is unmistakably clear is that his calculations gave him the basis for asserting a (possibly brand new) chronological system based on the life of Christ.

And the fact that Dionysius dared to propose this chronological system is perhaps again an indication that he was replacing the work of Victorius. For what he proposed was not only an improvement on the Paschal cycle of Victorius, but also an improvement in the chronology, despite his own apparent miscalculation. Victorius had constructed a chronology based on his calculation of Jesus' crucifixion and resurrection in AD 28—which incidentally implies that Jesus' birth took place about 4 BC at the latest, making Victorius actually closer than Dionysius to the actual historical occurrence of Christ's birth. By contrast, however, Dionysius' chronology began with his chosen year of the incarnation and birth of Christ, and was therefore a more accurate measure of the divine redemptive activity through the entire life of Christ.

But why did Dionysius ostensibly choose 25 December 1 BC as the birth date of Jesus? And why did he likewise choose 25 March AD 31 as the date of Jesus' resurrection? As we observed in previous chapters, these dates do not appear to have been chosen by any of the earlier Christian writers who commented on these matters, which leads us to the truth of Mosshammer's assertion that Dionysius stood rather by himself in his choice of dates.[25]

As an aside, Declercq has asserted that Dionysius actually meant to identify AD 1 as the year of Jesus' birth since 25 March of that year (and, according to Declercq, Dionysius' assumed date of the Incarnation) occurred on what would be the equivalent of Good Friday. In other words, 25 March AD 1 was the Friday just after the

---

[24] This is in agreement with Gustav Oppert, 116–117; F. K. Ginzel, *Handbuch der Mathematischen und Technischen Chronologie* (Leipzig: J. C. Hinrichs, 1906–1914), III, 179; Franz Rűhl, *Chronologie desMittelalters und der Neuzeit* (Berlin: Reuther and Reichard, 1897), 198; and Bruno Krusch, *Studien zur christlich-mittelalterlichen Chronologie* (Berlin: Akademie der Wissenschaften, 1938), 60.

[25] Mosshammer, 344.

vernal equinox, with the full moon occurring on 25–26 March, and the following Sunday (27 March) therefore being the proleptic equivalent of Easter.[26]

Declercq is undoubtedly correct in stating that Dionysius could not have missed this apparent agreement of astronomical phenomena. We should note two problems with this date, however: First of all, 25 March AD 1 corresponds to 11 Nisan 3761 in the Hebrew calendar, and therefore is impossibly early for correlation with Passover. Would Dionysius not have noticed this *non-correspondence* of dates as well? In addition, such a late date for the birth of Christ not only places it after the lifetime of Herod, and therefore in disagreement with the biblical narrative, but it also gives him a nearly impossibly short life and ministry. Specifically, 25 March AD 1 is exactly thirty years before 25 March AD 31. This would mean that Jesus' death occurred before he even reached his thirtieth birthday—which may strain the limits of even the putative one-year ministry of Jesus in the Synoptic gospels.

If, again, we wish to take Dionysius seriously as a man of faith, it must be admitted that he apparently adopted a curious mixture of gospel chronologies in determining the essential dates of Jesus' life. First of all, his apparent choice of 1 BC as the year of Jesus' birth and AD 31 as the year of Jesus' passion means that Dionysius adopted the shorter, one and one-half year chronology of Jesus' ministry as some have understood it from the Synoptic gospels. Declercq and Mosshammer have maintained that Dionysius would not have done this because of the chronicle written by Cassiodorus shortly before Dionysius' work that affirmed the longer chronology of John, which was also in agreement with the work of Prosper of Aquitaine seventy years previous.[27] The simple numbers of Dionysius' cycle, however, do not bear out this objection, and therefore point toward his adoption of the shorter chronology.

Secondly, and in contradistinction to his choice of chronologies for Jesus' public ministry, it is also apparent that Dionysius may have adopted the so-called Johannine chronology of Holy Week—the aforementioned supposition that Jesus' crucifixion took place on the

---

[26] Declercq, 145–147. Most other authors, however, have assumed the Dionysian date of Christ's birth to be 25 December 1 BC—just one week before the beginning of the year we know as AD 1.

[27] Declercq, 241; Mosshammer, 347.

day before Passover (13 Nisan), and that his resurrection occurred on 15 Nisan.[28] As noted, Dionysius' ostensibly chosen date for Jesus' resurrection is Sunday, 25 March AD 31. Rather problematically, however, this date happens to correspond to 13 Nisan 3791, not 15 Nisan. Before accusing Dionysius of miscalculation, however, it should be noted that the corresponding dates of 25 March AD 31 and 13 Nisan (or 15 Nisan) 3791 do not show up on his paschal cycle anyway. What does show up is the correspondence of the Easter date of 25 March AD 563 with 15 Nisan [4323].

The difference of two days between the dates of 13 Nisan 3791 and 15 Nisan 4323—which are exactly 532 years apart, and therefore should be the same date in both the Julian and Hebrew calendars—is explained by important calendrical developments which took place between the lifetime of Jesus and the writing of Dionysius. It should be remembered that following the Council of Nicaea in the early fourth century the vernal equinox was recalibrated to 21 March from 25 March, and that responsibility for determining the annual date for Easter appears to have been given to the astronomers at Alexandria. That much is clear from Dionysius' letter to Petronius. The recalibration was designed to correct the inevitable calendrical drift which occurred over the course of centuries with a Julian year which was slightly too long.

In spite of the four-day correction of the date of the vernal equinox in the fourth century, however, the Roman world was still using the Julian calendar, along with its flawed assumption of the length of the solar year at 365.25 days. As observed in chapter three, this assumption includes an overage of eleven minutes and fifteen seconds in every year, accumulating to just over three days in every 400 years. This means that by the lifetime of Dionysius—living and working a full two centuries after the recalibration to 21 March—the calendar was already in error by nearly two days. Furthermore, his calculation of the age of the moon at Easter for AD 563 was more than two centuries after the recalibration of the vernal equinox, with the result that the Hebrew calendar date for 25 March AD 563 had progressed two days past that of 25 March AD 31.[29]

---

[28] See chapter four for further discussion of the length of Jesus' ministry and the Hebrew dating of the events of Jesus' passion.

[29] This difference was discovered and noted with frustration by the Venerable Bede in his *De Temporum Ratione* two centuries later.

Therefore, the simplest explanation for Dionysius' apparent choice of 25 March AD 31 for the date of the resurrection of Jesus is that he did not notice this slight error in the Julian calendar. But what he did notice was the correspondence of 25 March AD 563 with 15 Nisan [4323], and assumed the same correspondence for 25 March AD 31 with 15 Nisan [3791]. Modern calculations, as discussed above, now show that this was in fact not the case, and that 25 March AD 31 actually corresponded to 13 Nisan. While this highlights another unfortunate "error" in the calculations of Dionysius, it also shows that he fully understood the structure and function of the 532-year grand paschal cycle. So although Rothwangl was roughly correct in asserting that Dionysius was focused on constructing an Easter cycle to fit his mathematical calculations, it had nothing to do with relieving his contemporaries of their eschatological fears with the approaching end of the 6,000-year World Week.[30] For Dionysius, the best-fitting Easter cycle was the correlation of Easter Sunday, 25 March AD 563 with his assumed *original* Sunday of Jesus' resurrection on 25 March AD 31. And his use of AD 532 as the starting point of the cycle shows his very obvious intent of furthering understanding, not only of calendrical cycles, but of the divine redemptive plan in the context of concrete human history.

INTERPRETIVE THOUGHTS ON DIONYSIUS

So goes the debate regarding possible dates for Jesus' conception, birth, death, and resurrection. While such speculation may appear unnecessary, unbiblical, or even laughable to some, it is certainly no more so than the rather awkward mathematics which appears to have been employed by chronologers such as Dionysius centuries ago.[31] Whatever the case, we have seen the multiple computistic difficulties of locating years in which all the solar and lunar phenomena coalesce to produce an ideal set of dates for the beginning and end of Jesus' life. This may also explain why Dionysius made no specific statement anywhere in his writings to specify whether Jesus' incarnation and birth were to be placed in 1 BC, AD 1, or any other year for that matter.

---

[30] See footnote 14 above.

[31] Borst, 723ff.

Consequently, we should exercise great caution in making conclusions beyond the obvious. Moreover, it should be noted that although several columns of Dionysius' Easter Cycle deal with lunar calculations—as do also some of the *Argumenta*—nowhere is there any obvious attempt to calculate the Hebrew (lunar) date or the Julian (solar) date of the incarnation, be it on 25 March 1 BC, or any other date. And although Declercq wisely quotes the Venerable Bede as an authoritative interpreter of the *Argumenta*, it is still not clear from the wording of the *Argumenta* themselves whether Dionysius' calculation of "which year it is since the incarnation of our Lord Jesus Christ" is meant to yield AD 1 or 1 BC. And if in our frustration we are tempted to view this as deliberate obfuscation on the part of Dionysius, then perhaps we are once again face-to-face with his mastery in mathematics and theology. In other words, perhaps the genius of Dionysius is shown not only by the details of his chart and the supporting explanations, but also by his deliberate omission of details regarding the specific year and date of Jesus' incarnation and birth.

Finally, we should beware lest we lose sight of the forest for the trees. In other words, we should never forget that Dionysius' original purpose was to calculate future dates for the celebration of Christ's resurrection from the dead, not his birth. As shown already in the previous discussion, celebration of the particular date of Jesus' birth lagged behind that of the debates concerning the date of his resurrection. In addition, we should remember that the inclusion of lunar cycles in these calculations has roots in pre-Christian pagan practices of the Roman world as much as in Judeo-Christian tradition. Although it is now surely clear to the reader where this author's sympathies lie, it must nevertheless be admitted that the question of whether astronomical cycles have any connection to the actual execution of the incarnation in the mind of God is entirely debatable. So although it is certain that valuable insight is to be gained from the debate over such calculations, it may be just as probable that this is nothing more than the kind of mystical parallelism of dates that fired the imaginations of ancient [and modern] computists.[32]

[32] Declercq, 146.

SO WHEN DID THE NEW MILLENNIUM BEGIN, ANYWAY?

Few people today outside a small circle of historians, theologians, and astronomers would know anything of Dionysius and his Paschal calculations were it not for the renewal of popular attention his work has received in the course of the last generation. For the question which suddenly gripped the popular mind in the late twentieth century was that of when the twentieth century was to end and the twenty-first to begin. As noted already, not everyone has appreciated the theological perspective and purpose behind the calculations of Dionysius' Paschal cycle and his consequent construction of the Christian, or modern western set of yearly numbers by which most of the world now dates itself. In addition, in recent generations numerous suggestions for possible improvements to the calendar have been advanced from a variety of scientific and political motivations. But whether Dionysius' theology has been appreciated or not, his widely-accepted chronology provided that irresistible question on millions of modern minds and tongues.

This author's first confrontation with that question occurred more than ten years before the turn of the century and millennium, in late 1989. A few people began asking out loud the question of whether the decade of the 1990s was to begin on 1 January 1990, or a year later. What occasioned such reflection among many people was the slowly-developing realization that they were about to enter not only the last decade of the century, but the last decade of an entire millennium. We were part of that generation who would step from one millennium into another. And although nearly everyone recognized that the numbers themselves and the transition from one era to another were arbitrary human constructs, at the same time many wondered whether there was something more cosmically significant to the moment.

Among some groups of Christians, of course, this meant looking for the imminent return of Christ or some related event. After all, the big change in numbers must mean something, right? In fact, that was the argument heard by this author in late 1989: "The big change in numbers happens at the beginning of 1990, not 1991. So I don't care what anyone else says!" While this writer did not bother to engage the debate at the time, something about that argument did not sound quite correct, although he had not yet researched the subject and hence could not argue it one way or another. He felt strangely like a Roman, with a preference for odd numbers that made him want to avoid beginning a decade with a year ending in "0." It was not long, however, before he began to discover Dionysius, who gave him the pleasant discovery that for one of the few times in his life his assumption was correct: The first year of every decade is the year ending in "1," which means that the same is true for the first year of every century and millennium. Consequently, the last year of every decade, and therefore of every century and millennium, is the year ending in "0."

Another giveaway to this chronological understanding—often missed by casual viewers—is found in the numbering of the centuries themselves. It is actually the final year of every century that gives the century its name. For instance, the first century of the Christian, or Common Era included the years 1–100, and so on. The century recently finished was called the twentieth century, even though the first ninety-nine years of that century all began with the prefix "19__." No matter, however, since it ended very appropriately with year 2000, or "twenty-hundred." After all, how could we call it the twentieth century if "20__" never showed up in it? Likewise, the current twenty-first century will extend through the end of the year 2100.

A prime reason for the omission of a year "0" between the BC and AD designations is clear from the numbers used in Dionysius' Paschal cycle and a host of other ancient Latin documents. Like many other ancient societies before them, the Romans used letters of their alphabet to express numbers and had no commonly-used symbol for zero. AD 599 in Dionysius' cycle is therefore represented by "DXCVIIII," AD 600 by "DC," and AD 601 by "DCI." This does not mean, however, that the Romans were incapable of zero-based calculations. An example is the third column of Dionysius' cycle, which lists the epact[33] as "nulla" (nothing, or zero) where appropriate at the beginning of every *decemnovennalis*, or 19-year cycle (such as in AD 513 and 532).

---

[33] Epact is a measure of the age of the moon on a given date. In Dionysius' cycle the epact is the age of the moon on March 22, just after the vernal equinox. Later calendars more commonly give the epact as the age of the moon at the beginning of the year. See http://www.webexhibits.org/calendars/calendar-christian-easter.html #anchor-epact.

Much of the argument concerning the omission of the year "0" must be asserted by inference, however, since Dionysius never stated specifically that by his calculation the incarnation and birth of Christ took place in 1 BC, or even in AD 1. Furthermore, as we have seen already, the designation of "AC" (Latin, *Ante Christum*), which most of us know as "BC" (*Before Christ*), made its appearance not with the writings of Dionysius but two centuries later with the *Ecclesiastical History of the English People* by the Venerable Bede.[34] It is Bede, therefore, who is to be credited with stretching Dionysius' numbering system backward to include the concept of time we know as *Before Christ*.

THE POST-DIONYSIAN LIFE OF THE DIONYSIAN PASCHAL CYCLE

Unfortunately, the age in which Dionysius lived and worked was not a peaceful, prosperous, and stable era known for scholarly productions. As noted at the beginning of this chapter, he lived at the beginning of what is often called the Dark Age(s), in just the first full generation after the decline and fall of Roman rule in the western Mediterranean. The last recognizable Roman ruler in Rome itself was the young Romulus Augustulus, who was deposed by Odoacer and his Ostrogoths in AD 476. After Odoacer came additional waves of Germanic invaders from northern Europe, and the once seemingly limitless territory and power of Rome were severely curtailed as the legions retreated eastward to defend their new capital at Constantinople.

The Christian Church was the one Roman social institution remaining in the west, and in time it was successful in converting the Germanic tribes not only to belief in Jesus but also to other elements of the Roman social and political system. But during the fifth and sixth centuries Italy was hardly a peaceful place for any lengthy period.[35] The legions under Belisarius re-established Roman/ Byzantine control of northern Italy and the entire Mediterranean

---

[34] Faith Wallis, *Bede: The Reckoning of Time* (Liverpool: Liverpool University Press, 1999), 156ff.

[35] I am touching only very briefly here on the migration and popularization of Dionysius' Paschal cycle. Engaging accounts offering more detail are given by Duncan, 104ff; Duncan Steel, 117ff.

region in the mid-sixth century, established a new working capital at Ravenna, and fulfilled Justinian's vision of a revived Roman Empire stretching from the Middle East to the Atlantic. Unfortunately, it lasted hardly a couple generations before the legions again retreated eastward and Italy was permanently out of "Roman" hands.

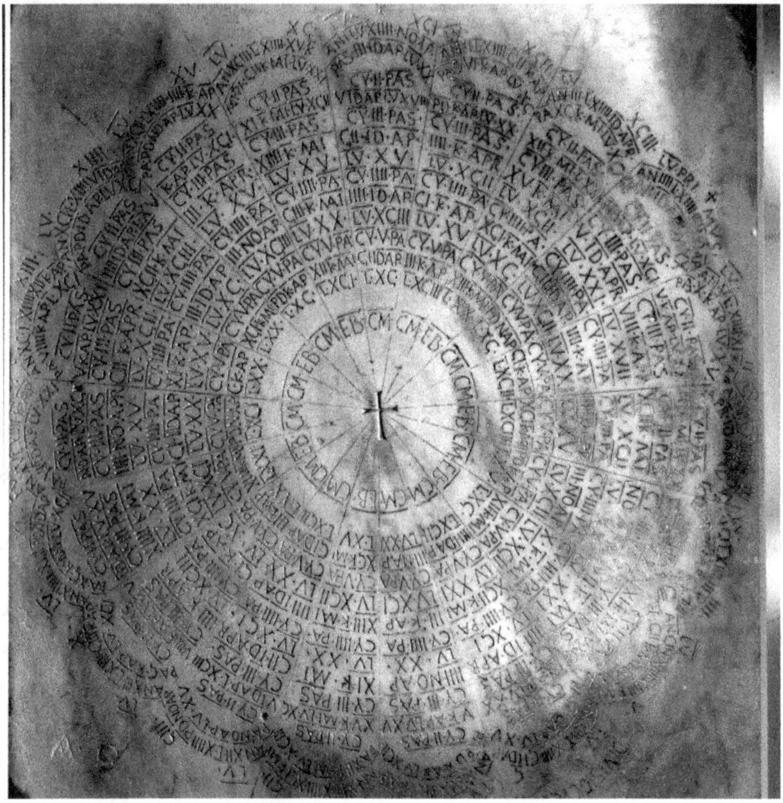

Figure 11. Paschal Cycle Calendar Stone. Archepiscopal Museum; Ravenna, Italy. Photo by Author. Used by permission of Opera di Religione della Diocesi di Ravenna.

In addition, the Byzantine victory in the western Mediterranean preceded the death of Dionysius Exiguus by no more than a few years. Nevertheless, a curious evidence of Dionysius' work which remains at Ravenna today in the Archiepiscopal Museum is an inscription of his Paschal cycle on a large marble slab. In a circular rosette style, it is divided into nineteen sections (resembling leaves/petals) for the nineteen years of the Metonic cycle. Flowing from the cross at the center, each set of years is designated as either "common" (having twelve lunar cycles) or "embolismic" (thirteen lunar cycles). Then beginning from the outer edge, each section lists the five years of Dionysius' cycle which correspond to that stage of the Metonic cycle, and the date of Easter for each year. Likewise included are the dates of Passover and the age of the moon on Easter Sunday. While little is known about the history of the stone itself, its very limitation to the ninety-five years of Dionysius' cycle is a sure indication that it was inscribed shortly after his creation of the cycle itself. In addition, the impressive physical size of this calendar (approximately three-feet square) is very likely an indication of its importance—that it was displayed prominently for use by ecclesiastical authorities.[36]

The temporary victory of the Byzantines in northern Italy ironically forced the flight of one of the best scholars of the period away from the official Roman dominions—the aforementioned Cassiodorus (ca. AD 505–585)—for he had worked for the Ostrogoths. Whether Cassiodorus was personally acquainted with Dionysius is not known. But what is known is that Cassiodorus took a copy of Dionysius' Easter cycle—presumably much lighter and more portable than the marble slab—when he fled from the Byzantines and established a monastery at Squillace in the far south of Italy, on land owned by his family. A generation later, after the death of Cassiodorus and when Dionysius' cycle was nearing its end, the aforementioned Felix Gillitanus—a student of Cassiodorus—prepared a new Easter cycle for AD 627–721. The cycle created by Gillitanus apparently made its way to Rome, where it was copied rather extensively and sent out to local dioceses.

One of those places where Gillitanus' work showed up and became a matter of sharp controversy was at the far western reaches of the former Roman domain. The British Isles were not immune to the ravages of the barbarian hordes that had displaced the Romans as lords of the west. The Jutes, Angles, and Saxons all did their share of

[36] Little information on the history of the stone is given, except that the slab came from a quarry on the island of Proconnesus in the Sea of Marmara, near the coast of modern Turkey. One travel blog described this stone as "striking but indecipherable" (http://www.travel-wonders.com/2009/07/touring-mosaic-city-ravenna-italy-part_10.html). It is certainly striking, but far from indecipherable, as many of this writer's students will attest.

damage as invaders, and eventually settled down to administer their newly-conquered dominions. For the next few centuries—in the age of the legendary King Arthur and his contemporaries—England was nothing more than a patchwork quilt of competing and sometimes warring kingdoms. At the periphery of these events were Ireland and Scotland, who largely escaped the devastations of the barbarian invasions and therefore did not experience the same depth of "Dark Age" seen elsewhere. In fact, the success of the Gospel in Ireland, beginning with the ministry of Patrick (AD 387–461), was such that it became a center of scholarship over the next two centuries. And that scholarship included the use of the same old and inferior Roman computus which had informed the cycle of Victorius.[37]

So the Paschal cycle of Dionysius, which was based on the admittedly superior Alexandrian lunar calculations and was apparently adopted by Rome, was lucky to survive the barbarian onslaughts following the Roman decline in the West. Then Alexandria itself was conquered with the advance of Islam in the seventh century, and the Christians soon became a small and second-class minority. Meanwhile, the 84-year Roman Paschal cycle, with the same lunar measurements used by Victorius, was transported to Roman Britain and survived rather unscathed. The Scots-Irish brand of Christianity was quite successful until Rome sought to re-establish her norms in the British Isles in the mid-seventh century. At the Synod of Whitby in AD 664 the competing Paschal cycles confronted one another once more, and once more the Alexandrian/Dionysian work was judged superior, although it appears that the deciding factors at this synod very possibly had more to do with politics and personality than with sound scholarship.[38]

Six decades later in Britain the Dionysian cycle was utilized and commended by the Venerable Bede (672–735) who, as we have seen, extended its numbers backward to include those years we now call "BC," or "Before Christ." Just before the end of that same century the scholarship of the British Isles flowed back to the European continent when Alcuin (735–804), Archbishop of York and a renowned

[37] For more on the success of the Christian Gospel and the flowering of Christian scholarship in Ireland in the otherwise "Dark Age," see the engaging account of Thomas Cahill, *How the Irish Saved Civilization* (New York: Anchor Books, 1996).

[38] Steel describes the alleged debates and traditional historical accounts of the Synod of Whitby as nothing less than "subterfuge" (122ff).

educator, was persuaded by Charlemagne to become master of his palace school. Within a few more years the military successes of Charlemagne brought nearly the entirety of western Europe under a single political regime for the first time in over three hundred years, and on Christmas Day in AD 800 he was crowned by Pope Leo III as the first Holy Roman Emperor. Consequently, the short-lived but glorious Carolingian Renaissance succeeded in copying and preserving many priceless works of antiquity.

Although the Dionysian Paschal cycle and the *Anno Domini* chronological orientation were still not held universally, its adoption by the Roman Church and the Holy Roman Emperors assured its survival and eventual growth to prominence. The darkest period in the cultural history of the west was over, and one of the few lights which had shone through it and kept the fires of scholarship burning was the schedule of Easter celebrations penned by an otherwise unknown monk. "Little Dennis" would have probably preferred it that way.

6

URBAN'S BIG MISTAKE, GREGORY'S BIG FIX, AND A SIMPLE LINE

THE CHURCH AND TIME MEASUREMENT DURING THE SCIENTIFIC REVOLUTION

I, Galileo Galilei, son of the late Vincenzo Galileo of Florence, in the seventieth year of age, having been brought personally for judgment and kneeling before you, eminent and most Reverend Cardinals, and General Inquisitors of the Christian Republic against heretical depravity, and having before my eyes and touching with my own hands the holy Gospels, swear that I have always believed, believe now, and with the help of God will for the future believe everything that is held, taught, and believed by the holy Catholic and Apostolic Church. But whereas I have been enjoined by this Holy Office to abandon the false opinion that the sun is the center of the world and that it does not move and that the earth is not the center of the world and that it moves, and that I am not to defend in any mode, whether in voice or in writing, this false doctrine, and further that this doctrine is contrary to Sacred Scripture, after having spoken and written about it with many efficacious arguments, without bringing any solutions, and therefore being under vehement suspicion of heresy, namely, of having taught and believed that the sun is the center of the world and immobile and that the earth is not the center and that it moves;

> Therefore, wishing to remove from your Eminences and from every faithful Christian this vehement suspicion, rightly conceived against me, with sincere heart and unfeigned faith I abjure, curse, and detest the aforesaid errors and heresies, and generally each and every other error, heresy, and sect contrary to the Holy Church; and swear for the future not to say or assert, in voice or in writing anything that may give rise to a similar suspicion about me, and that if I know a heretic or have suspicion of heresy will denounce them to this office, or to the Inquisitor or Ordinary wherever I may be. . . .
>
> Moreover, I swear that I shall fulfill and observe all the penances which have been laid upon me by this Holy Office. And if I violate any of my penances and oaths (which God forbid!), I will submit myself to all the punishments which have been decreed by this Holy Office. So may God help me, and his Holy Gospels, which I touch with my own hands.[1]

Thus concluded one of the most negative and tragic chapters in the history of the relationship between Christianity and scientific endeavors. Unfortunately, it is also one of the best-known chapters in that history. Since especially the works of chemist John William Draper (1811–1882) and Cornell University founding president Andrew Dickson White (1832–1918), both of whom sought deliberately to cast this relationship in historically antagonistic terms[2], Christianity in general and the Roman Catholic Church in particular

---

[1] Galileo Galilei, *Opere* (National Edition. Antonio Favaro, ed. Florence: Barbè1890–1909), XIX, 406–407. Author's translation. Also available at http://pinakes .imiss.fi.it:8080/pinakestext/annotate.seam?conversationID=6130. Documents and commentary on the trial of Galileo are found in Maurice A. Finocchiaro, ed., *The Galileo Affair: A Documentary History.* Berkeley: University of California Press, 1989), 292–293.

[2] John William Draper, *History of the Conflict Between Religion and Science* (Cambridge: Cambridge University Press, 2009 [1874]); Andrew Dickson White, *A History of the Warfare of Science with Theology in Christendom* (2 vols. New York: D. Appleton, 1896). Analysis of the influence of the Draper-White "conflict thesis" and the evolving relationship between religion and science over the past few centuries is found in David C. Lindberg and Ronald L. Numbers, "Introduction," *God and Nature: Historical Essays on the Encounter Between Christianity and Science* (Lindberg and Numbers, eds. Berkeley: University of California Press, 1986), 1–18 Since the late twentieth century the conflict thesis has been largely rejected by academicians and serious historians, although it has lingered in the popular media.

have been cast popularly in the mold of an anachronistic bastion of fear and ignorance over against the heroic and liberating progress of modern science. Draper and White were fully in agreement with the *Zeitgeist* of the late nineteenth century, of course, including the concurrent rise of Darwinian naturalism, which offered a convenient creation story for moderns who were uncomfortable with the moral obligations brought by belief in the Judeo-Christian God. For authors like Draper and White, the trial of Galileo (1564–1642) has been often viewed as Exhibit A in the popular indictment against all organized religion as little more than outmoded and oppressive superstition, even in spite of more recent inquiries by the papal office and the Pontifical Academy of Science aimed toward the effective rehabilitation of Galileo.

FIRST, URBAN'S BIG MISTAKE

Some details often left untold in the popular histories and therefore missed by the casual reader, however, make this a more complicated and in fact a more interesting story. As implied in Galileo's statement above, his trial in 1633 was actually the second time he had been questioned by the Inquisition concerning the content of his teaching. At the first inquiry in 1616 he had been admonished by Grand Inquisitor and Cardinal Robert Bellarmine (1542–1621) to teach and write of the heliocentric theory only as theory, not as fact, since heliocentrism had not yet been proven[3] and was apparently contrary to plain statements in Scripture. To this Galileo had agreed, and indeed assumed that he was abiding by this restriction in his *Dialogue Concerning the Two World Systems* (1632) when he posed the matter as a friendly debate between proponents of geocentrism and heliocentrism. Like most academicians and scientists of his day—and especially in his homeland of Italy, where the political and social influence of the Roman Church was still abundantly apparent—Galileo was somewhat dependent upon ecclesiastical patronage for his income and the publication of his works. Moreover, his *Dialogue* had already been given implicit approval for publication by the

[3] The fact that no convincing scientific proofs of heliocentrism had yet been given by 1633 was advanced by recent statements of the Pontifical Academy of Science as a substantive, if rather unsatisfying argument in defense of the Inquisition's case against Galileo. See Maurice A. Finocchiaro, *Re-Trying Galileo* (Ewing, NJ: University of California Press, 2005), 347ff.

Inquisition's representative in his hometown of Florence. So being faced with the charge of teaching heresy was as much a surprise to Galileo as to anyone else.[4]

And if the charge itself was a surprise, the outcome of the trial was even more so. During the proceedings Galileo produced a letter received from Bellarmine shortly after their 1616 meeting which confirmed the content of their conversations and the admonition given. But he was countered by the inquisitors, who produced another letter, allegedly also written by Bellarmine, in which he was specifically commanded to cease teaching and writing on Copernicanism altogether. The latter letter is now regarded as a forgery, but since Bellarmine had already died years before the trial there was no way to question its authenticity. Since Galileo was therefore found in disagreement with the Church on a matter of biblical interpretation, he risked condemnation and punishment as a heretic. His only recourse was to construct a letter of confession and recantation. With the losses of members, funds, and prestige which the Roman church had already suffered since the beginning of the Protestant Reformation a century earlier, and the ongoing tensions between the several nations involved in the Thirty Years' War (1618–1648), there was little chance that even the personal affection for Galileo by Pope Urban VIII (1568–1644; accession 1623) could have dissuaded an Inquisition which was hungry for a victory over a perceived enemy.

So at least in this one well-known case, and perhaps in one other[5], the critics of organized Christianity and especially of its allegedly

---

[4] For the documents concerning Galileo's trial and the several subsequent reviews of his case, see Finocchiaro, *The Galileo Affair*; Finocchiaro, *Retrying Galileo.* Analysis of the trial and its historical setting is found in Jonathan Kirsch, *The Grand Inquisitor's Manual: A History of Terror in the Name of God* (New York: HarperOne, 2009), 161ff; William R. Shea, "Galileo and the Church," *God and Nature*, 114ff.

[5] That other case would be the Scopes "Monkey Trial" in Dayton, TN during the Summer of 1925. John Scopes, a high school biology teacher, was enlisted by the American Civil Liberties Union to be their test case in challenging a recently-passed Tennessee law which prohibited the teaching of evolution in the public school classroom. The trial became a media circus because of press coverage and the involvement of famous personalities such as William Jennings Bryan as counsel for the prosecution and Clarence Darrow as counsel for the defense. Although Scopes was convicted of violating the law and fined $100, it was later overturned on appeal. The trial was later made into a Broadway play and Hollywood film with the creation

anti-progressive influence on human culture appear to have a valid argument. But can such an argument be maintained consistently when one observes the many places and occasions of engagement between the Christian faith and scientific research over the past several centuries? The popular perception of religious faith, and especially the Christian faith, as the enemy of modern science has persisted in modern western culture in spite of the abandonment of the Draper-White thesis by most historians since the mid-twentieth century. Even more significantly, the case of Galileo was given a re-evaluation by the Pontifical Academy of Sciences beginning in 1979. In his concluding address on the matter in October 1992, Pope John Paul II admitted that in 1633 most theologians of the Church were not prepared to distinguish academically between the Bible and its interpretation, which regrettably allowed a matter of scientific inquiry to become in addition a matter of doctrine.[6] Unfortunately, while the findings of the Academy effectively rehabilitated Galileo in the eyes of the Catholic Church, this action was viewed by many as an attempt by the Church to rehabilitate its own image before a modern, scientifically-minded world.

As captivating as the story of Galileo Galilei may be, however, this chapter is not about him, his scientific discoveries, or the Inquisition. Nor is this an attempt to defend the Roman Church's actions in the Galileo affair, which would be an unnecessary departure from our historical investigation of Christianity and chronology. Rather, it is the contention of this chapter that when one observes in this same era of the Scientific Revolution the high level of investment made especially by the Roman Church toward the

---

of *Inherit the Wind*, in which Bible-believing Christians were mercilessly pilloried as narrow-minded, anti-intellectual, Bible-thumping bigots. See George M. Marsden, *Fundamentalism and American Culture* (New York: Oxford, 1980); Joel A. Carpenter, *Revive Us Again: The Reawakening of American Fundamentalism* (New York: Oxford, 1999); David Goetz, "The Monkey Trial," *Christian History* 55 (July 1997), http://www.christianitytoday.com/ch/1997/issue55/55h010.html. For transcripts and newspaper accounts from the trial, see http://www.law.umkc.edu /faculty/projects/ftrials/scopes/scopes.htm.

[6] "Ansprache von Johannes Paul II. An die Teilnehmer der Vollversammlung der Päpstlishen Akademie der Wissenschaften. 31.October 1992." Author's translation. http://www.vatican.va/holy_father/john_paul_ii/speeches/1992/october/documents/hf_jp-ii_spe_19921031_accademia-scienze_ge.html. See also Steven Gertz, "Upon Further Review . . . : 360 Years Later, the Pope Admits the Church's 'Error of Judgment' toward Galileo" *Christian History* 76 (Vol. XXI, No. 4; 2002), 17.

accuracy of time measurement, the Galileo debacle stands out as the exception rather than the rule.

GREGORY'S BIG FIX

Although the proclamation of the correction to the Julian calendar by Pope Gregory XIII took place a full half-century before the trial of Galileo, when viewed with the benefit of our longer historical lens of more than four hundred years later it becomes apparent that Gregory's calendrical revision of 1582 was actually more indicative of the stance taken toward scientific endeavors by Christianity in general and the Roman Catholic Church in particular. Seen from the larger historical perspective, the calendrical revision was a very good thing for the legacy of Gregory as well, who is otherwise known for having celebrated the "Saint Bartholomew's Day Massacre" of nearly 30,000 Protestants in France a decade earlier by commissioning a commemorative medal in honor of the event.[7]

Although Gregory gets the credit for the calendar which has ever since borne his name, the hard work of correcting and constructing the calendar was accomplished by several lesser-knowns, especially by Aloisius Lilius (ca. 1510–1576) and Christopher Clavius (1538–1612). The need for reform of the Julian calendar had been known among astronomers for centuries, yet no concrete action was taken until a recommendation to the papal office on this matter was made by the reforming Council of Trent. Once again, awareness of this development offers a slightly more positive reading of an event in Catholic history which is otherwise known to non-Catholics as merely

---

[7] Understandably, if unconvincingly, Catholic authors have attempted a defense of Gregory's apparent celebrations following the massacre of Huguenots in France in August 1572 by claiming that the Pope knew nothing of the horrific details, that he wept openly upon hearing of the details, and that he was actually celebrating the demise of an alleged Huguenot conspiracy to overthrow the French monarchy. See Michael Ott, "Pope Gregory XIII," *The Catholic Encyclopedia* (New York: Robert Appleton, 1910), http://www.newadvent.org/cathen/07001b.htm. Concerning the commemorative medal mentioned above, the image on the medal's reverse of an angel holding a Latin cross in one hand and a sword in the other pointing toward dead bodies lying on the ground would seem to indicate celebration of the event by the papacy rather than mourning. See also Carter Lindberg, *The European Reformations* (New York: Wiley-Blackwell, 1996), 295; Niall Kilkenny, "The Saint Bartholomew's Day Massacre," http://www.reformation.org/saint-barts-day-massacre.html.

a negative reaction to the challenges of the Protestant Reformers, to whom by this time the Roman Church had already lost sway over millions of souls and square miles of territory. It should also be noted that at an early stage at Trent—conducted in three major sessions stretching from 1545–1563—official approval and support had also been given to a new monastic order, the Jesuits. These zealous soldiers of Rome thenceforth became a primary weapon in the Church's efforts to not only win back the hearts and minds of Europeans, but also to extend the Roman Catholic Gospel through the mercantile alliances of Portugal, Spain, and France to previously unevangelized regions of Africa, Asia, and the newly-discovered lands of the Western Hemisphere.

Why a reform of the Julian calendar? As noted previously, the Julian calendar contained a small error in its calculation of the solar cycle, or year. The solar year—by which we normally mean the Tropical Year, is actually eleven minutes and fifteen seconds short of the Julian estimate of 365.25 days. While this shortage is insignificant in the course of a single year, when compounded over generations it becomes approximately three days and three hours in the course of 400 years. Calendrically speaking, it means that the actual occurrence of the vernal equinox slips slowly backward or earlier relative to the numerical dates on the calendar.

For most Europeans in the sixteenth century, the vernal equinox was still the traditional beginning of a new year and therefore the time of annual festivals. In most places these ran for an *octave*, or period of eight days from 25 March (the original date of the Julian vernal equinox) through 1 April. Recognition of the vernal equinox at 21 March had been accomplished by the ancient astronomers at Alexandria, whose measurements had apparently informed the Council of Nicaea in the early fourth century[8], although no one had attempted to re-orient observances at the popular level. But with the Julian error left uncorrected, by the late sixteenth century it was apparent to astronomers that the vernal equinox had slipped back by at least ten more days. The calendar itself was thus out of sync with

[8] This is evident in the annual "Festal Letters" of Athanasius, bishop of Alexandria (AD 328–373), who allowed for Paschal occurrences as early as 22 March. See *Nicene and Post-Nicene Fathers* (Second Series. Grand Rapids: Eerdmans, 1991 [1890]) IV, 506ff.

astronomical time, and spring and every other season were becoming progressively earlier with each year.

This brought to the fore once again the difficulties with the multiple factors involved in the proper scheduling of Easter which were discussed in the previous two chapters: Easter was understood to be the first Sunday after the first full moon after the vernal equinox. The baseline, then, for such calculations was the assumed vernal equinox on or very near 21 March. But if 21 March was no longer the true vernal equinox, the whole system of dates was subject to re-calculation. And the dates would continue to be subject to re-calculation once every century or so as long as the calendar remained uncorrected.

Therefore, following the recommendations made by Lilius to a papal commission in 1575, which were then refined and supplemented by Clavius in 1581, Pope Gregory proclaimed his *Inter Gravissimus* on 24 February 1582. The most immediate concern was the correction of the ten-day overage accumulated by the calendar since Nicaea, which was effected by eliminating ten days from the following October. It was therefore proclaimed that the day immediately following 4 October 1582 was to be regarded as 15 October, not 5 October. (Pity anyone having a birthday on one of those days!) Secondly, since the overage of eleven minutes and fifteen seconds each year accumulated to just over three days in the course of four hundred years, it was proclaimed that no leap year day would be inserted in the centennial years (those years at the end of every century ending in "__00") unless that year were divisible by 400. That meant the approaching end of their own sixteenth century (AD 1600) would indeed be a leap year, since 1,600 is divisible by 400. It would appear that they left it to a later generation to remember that the successive centennial years of 1700, 1800, and 1900 were not to be regarded as leap years, in order to keep the ancient error from creeping back into the calendar.[9]

As we should only expect in this era of sharp political rivalries, and perhaps even sharper religious rivalries, it was inevitable that not everyone would immediately agree with and apply the Gregorian reform. As expected, Catholic bishops and local clergy complied with little delay, so that their observance of the seasons of the liturgical

---

[9] For the text of Gregory's *Inter Gravissimus* in Latin, French, and English, see http://www.bluewaterarts.com/calendar/NewInterGravissimas.htm.

calendar and church holidays would not be out of step with Rome. Catholic monarchs in Spain, Portugal, Poland, and most of Italy likewise complied in adjusting their civil calendars. They were quickly followed by France, and within the year by the Northern (Protestant) Dutch provinces. The Southern (Catholic) Dutch provinces (modern Belgium) complied in 1583.

In most Protestant regions, understandably, the new calendar was met with suspicion as a kind of papal imposition on lands where his decrees no longer held sway. Yet because Protestants generally welcomed the Copernican astronomical system and other scientific findings of the era, it was not long until this became viewed as a non-theological matter. Nevertheless, it was not until 1700 that the Gregorian system was adopted in Denmark (including Norway) and most German Protestant states. England and its colonial dominions made the switch in 1752—by which time it was necessary to eliminate eleven days—and Sweden and Finland complied in 1753. Finally, it should be noted that in the Orthodox dominions farther east the transition was even longer in coming, with Russia adopting the Gregorian calendar in 1918 after the Bolshevik Revolution, and Greece coming into compliance in 1923.[10]

Speaking of the Orthodox dominions, an unintended consequence of Gregory's reform was to furnish one more reason for alienation between Eastern and Western Christians. Officially separated since the mutual condemnations of 1054, the papal representatives sent to Constantinople in the wake of Gregory's proclamation could hardly have entertained much optimism over the reception their new calendar would likely encounter. Predictably, although their report was received graciously, and even though Orthodox theologians were equally aware of the need for calendrical reform, Gregory's calendar was rejected because it would require altering the traditionally understood method for calculating the date of Easter.[11]

---

[10] For a more detailed story of the adoption of Gregory's calendar, including regions outside Europe and its dominions, see Duncan, 288ff.

[11] Even more than three centuries later, when the Orthodox religious calendar was adjusted to agree with the Gregorian calendar by a reforming council in May 1923, many Russian priests discovered to their dismay that it motivated full-scale parish revolts—perhaps a reaction to the social and political makeover of Russian life by the Bolsheviks. Whatever the reason, the reform was short-lived. Consequently, most Russian Orthodox believers still follow the Julian calendar for ecclesiastical purposes, while operating on the Gregorian calendar outside the church. See Marcos

Furthermore, eastern Christians had attempted to maintain the practice of ensuring that Easter always occurred after Passover—a practice which had not been maintained in the west due to the shifts in Jewish calendrics discussed in previous chapters and the smaller numbers of Jews living in the West. They were quick to note that Gregory's new calculations took no account of Passover, and hence allowed Easter to fall as early as 15 Nisan on the Hebrew calendar, which was unacceptably early for Orthodox believers. Consequently, ever since the implementation of the Gregorian reform there have been two distinct dates of celebration of Christ's resurrection on the worldwide Christian calendar. Only on the average of about once every three to four years do both eastern and western Christians celebrate on the same day. It could even be said that there are three Easters in a given year, due to the fact that those Orthodox communions who have officially adopted the Gregorian calendar still celebrate the resurrection on the same day as their counterparts using the Julian calendar (and are therefore out of step with western Christians), but use Gregorian dates for the occasion (and are therefore out of calendrical step with the rest of their Orthodox brethren).[12] So while the wording of Gregory's proclamation makes it clear that this action was intended as an act of faithfulness to the words of scripture and apostolic practice—and therefore much more than a mere improvement of the calendar—it was clear that not all Christians agreed that this was the correct way to perceive the hand of God in human history.

A SIMPLE LINE

Gregory's calendrical reform is certain evidence that the patronage of at least this one scientific endeavor by the Catholic Church was motivated by a sizeable element of self-interest—the accurate

---

J. Montes, "The Orthodox Ecclesiastical Calendar," http://www.smart.net/~mmontes /ortheast.html.

[12] See Appendix G, "Dates of Easter, 1956–2200," for a comparison of Western, Orthodox (Julian), and Orthodox (Gregorian) dates for Easter. Also illustrated by this comparison of eastern and western dates for Easter is that the Gregorian reform occasioned an adjustment in the succession of embolismic years in a given Metonic cycle. While the Orthodox dates for Easter still include embolisms in years 3, 6, 8, 11, 14, 17, and 19 of each cycle, since the Gregorian correction the western dates for Easter include embolisms in years 1, 4, 6, 9, 12, 15, and 17.

scheduling of Easter according to both solar and lunar requirements. After all, how could the church claim to be the repository of spiritual truth and the pope claim to be the Vicar of Christ if they could not accurately regulate their own celebration of what they regarded as the most important event in human history—the resurrection of Christ from the dead? Moreover, it should be noted that during this era members of Catholic religious orders were conspicuous for their numbers involved in scientific research—especially in the case of the Jesuits.[13] Their pursuits ran the gamut of scientific endeavors in fields both integral and peripheral to specifically Christian concerns—astronomy, biology, medicine, cartography, meteorology, and physics.

Self-interest or not, however, the Catholic Church continued to be as much a patron of science in the generations following the trial of Galileo as they had been in those preceding it. In fact, it might even be asserted that in the church's continuing patronage of science after the trial of Galileo one can perceive a degree of self-imposed penance for their mishandling of the affair. Whatever the case, the ongoing interest of the church in astronomical and calendrical accuracy is seen nowhere more visibly than in the construction of *meridiane* in notable church buildings in Italy during the one and one-half centuries before Galileo's trial, and more during the century following the trial.

A *meridiana* is at its core a simple line which performs a function inside a building similar to that accomplished by a sundial outside a building. It tells time, yet not in the same way. Rather than measuring the changing angle of the sun in its daily passage across the heavens from east to west, a meridiana is a line oriented north-south to note the sun's changing angle toward a particular location at a particular time each day. This angle changes as the Earth proceeds through its annual orbit and its inclination of nearly 23.5° on its axis causes the variations of light and temperature, as described in chapter three. In a region such as Italy, which is north of the Tropic of Cancer and

---

[13] Ashworth asserts that, in spite of the large numbers of Jesuits involved in scientific endeavors, in the end they contributed only modestly to the advancement of scientific knowledge. First, they were overly eclectic, and therefore failed to discriminate between relevant and irrelevant sources of information. Secondly, they viewed nature as emblematic—symbolic of deeper spiritual meanings—and therefore in their eclecticism lacked a philosophical superstructure to bind their findings together. See William B. Ashworth, Jr., "Catholicism and Early Modern Science," *God and Nature*, 153ff.

therefore never experiences a sun directly overhead, a carefully constructed meridiana is therefore an accurate way to measure the sun's changing angle through the course of the year.

To produce the right conditions for accurate measurements via a meridiana, Italy's large church buildings provided the perfect locations. The first requirement was a large south-facing wall with a window or another opening high enough to have unobstructed exposure to the sun. Secondly, it was required that a small hole, or *gnomon*, be cut through the south wall or a window high on the wall at an angle suitable for casting a beam of sunlight directly on the line below. Thirdly, the building needed a north-south expanse long enough to accommodate a line cut into the floor whose northern terminus (on which the sun would shine at its lowest angle at the winter solstice) was still on the floor and not creeping up the north wall. Since the more common orientation of Christian church buildings is east-west—so that worshipers face eastward toward Jerusalem—only very large buildings with a long north-south expanse were suitable. Alternatively, there were some church buildings with an unorthodox north-south orientation.

The earliest and shortest of the meridiane was constructed at perhaps the busiest and best-known of their locations.[14] About 1475 a line of white marble was inserted into the floor in the north transept of the cathedral church of Santa Maria del Fiore—better known as the Duomo, or cathedral—in Florence. The line was illuminated by sunlight through a gnomon in the south side of the cavernous dome. Its architect was Paolo del Pozzo Toscanelli[15] (1397–1482), one of

---

[14] Only a brief description of each of several meridiane and historical details of their construction are given herein. The reader with further interest in this topic is encouraged to consult J. L. Heilbron, *The Sun in the Church: Cathedrals as Solar Observatories* (Cambridge: Harvard University Press, 1999), 62ff. Although many pages of Heilbron's presentation are thick with geometrical and astronomical detail, his wide-ranging historical narrative is nevertheless quite engaging.

[15] Toscanelli is also known as a motivating source behind Christopher Columbus' ambition to reach East Asia by sailing Westward across the Atlantic. Columbus heard of a 1474 letter Toscanelli had sent to Fernão Martins, which was then delivered to King Alfonso V of Portugal, in which Toscanelli claimed that the Earth was less than 20,000 miles in circumference. The original was lost, but about 1481 Toscanelli transcribed it and sent it along with a map to Columbus, who then took it on his voyage in 1492. If not for Toscanelli's apparent miscalculation of our planet's size, Columbus may not have become the "discoverer" of the Western Hemisphere and

the leading thinkers of the early Italian Renaissance and a friend of Filippo Brunelleschi (1377–1446), designer of the dome. Fortunately for Toscanelli, Brunelleschi could object to the project only silently and posthumously from his crypt below as the hole forming the gnomon was cut into his masterpiece above.

A second and longer meridiana for Florence was installed a century later at the nearby convent church of Santa Maria Novella. Its architect was Egnazio Danti (1536–1586), for whom the church building served as an astronomical laboratory. A godsend for Danti was the large south-facing entrance—a byproduct of a remodeling and expansion project begun in 1279 which re-oriented the main axis of the sanctuary from east-west to north-south. Danti's first astronomical addition to the exterior of the southern façade was an astronomical quadrant in 1572. The quadrant is a large square of marble attached to the wall on one side, with lines of sight showing the angle of the sun at the extremes of summer and winter solstice, as well as in the center at both equinoctia. The quadrant was followed by an armillary sphere in 1574. The sphere is situated so that all its interior exposed surfaces are completely dark at noon only twice per year—at both vernal and autumnal equinoctia. In addition, the position of the sphere's shadow on the wall assured accurate dating of the equinox.

Unfortunately for Danti, his construction of the meridiana inside the church and the gnomon in the huge rose window on the south wall were only partially finished when he was forced to leave Florence in late 1575 on moral charges which have never been clarified. Even more unfortunate is the fact that Danti's meridiana was eliminated in a renovation of the interior begun in 1858, although the small hole in the window above remains. It is possible that Danti was the victim of changing political structures in Florence after the death of his patron Cosimo I de Medici and the accession of Cosimo's son Francesco. Whatever the case, it seems that he had little trouble finding employment or patrons, for by the end of the same year he was

---

Spain might not have become a colonial and commercial superpower. See Hugh Thomas, *Rivers of Gold* (New York: Random House, 2005), 56–59.

Incidentally, it should be noted that although Toscanelli was widely criticized, the actual circumference of the earth at 37° North Latitude (the level of southern Spain and Sicily) is just over 20,000 miles. The circumference at the equator is 25,000 miles, as asserted by Toscanelli's opponents. See Sigurd Humerfelt, "Earth according to WGS 84," http://home.online.no/~sigurdhu /Grid_1deg.htm.

working in Bologna, and the next year was appointed professor of mathematics at the university.

It is in Bologna that the longest meridiana ever created was implanted into a church building. The Basilica Church of Bologna's patron San Petronio was designed from its beginning in AD 1390 to be a statement of the city's spiritual, economic, and political vitality. In fact, the early ambition of the city's patrons was to exceed the size of the Cathedral of Saint Peter in Rome. The unfinished front façade of San Petronio is still today a witness to the fact that even before Danti's arrival in 1575 papal pressure on the city fathers had succeeded in convincing them to divert funds from the church to the university and other civic projects.

Danti obtained the necessary permissions and began construction of the meridiana immediately, finishing it before the end of his first year at Bologna. Taking advantage of the nearly astronomically correct north-south orientation of the building (it was constructed to face the Piazza Maggiore, the central square of the city, rather than to conform to astronomical specifications), Danti cut a hole in the rear wall of the building which allowed a long beam of sunlight to be cast down a side aisle. The meridiana was placed carefully so that the sun's image would be centered on the line at the astronomical noon of the day. A longer meridiana allowed for more precise measurements, which was to prove instrumental in several astronomical debates not only during Danti's lifetime but over the next century which followed.

The excellence of Danti's work in Florence and Bologna attracted the attention of the aforementioned Pope Gregory XIII, who appointed him to the commission charged with developing the new calendar. Relocating to Rome about 1580, he immediately went to work constructing a meridiana in the Torre dei Venti, a private chapel of the Pope located in the Vatican. Although Danti played only a minor role as a member of the Pope's commission, it is perhaps a tribute to his industriousness and long-lasting influence that an uncorroborated story of the period has it that Gregory was finally convinced to propagate the new calendar in 1581 after Danti used the new meridiana to show him that the vernal equinox had regressed by ten days since Nicaea to 11 March.

Nearly seventy years after Danti's death his work was honored once more and extended by one of the brilliant minds of that generation. During the last flurry of construction on San Petronio it was decided that the south wall should be demolished and replaced as the building was lengthened. This meant the end of the gnomon, or "eye" which allowed sunlight to shine on Danti's meridiana. But rather than allow the line to be eliminated or even shortened—which would have diminished its accuracy—Giovanni Domenico Cassini (1625–1712) came to its rescue. More than just a rescue, however, Cassini succeeded in gaining permission to cut a hole through the ceiling and roof, even higher than the original gnomon, and to lengthen the line on the floor. Finished in 1655, the new eye at 1,000 inches above the floor (27.07 metres) and the new line at nearly sixty-seven metres in length are still a marvelous sight to behold even today.

Although Cassini left Bologna in 1669 to pursue even larger projects as chief astronomer at the Observatoire Royal in Paris, he returned to visit his Italian masterpiece with regularity—most notably when he returned to check its accuracy forty years after its construction in 1695. One of the wonders of the line itself was Cassini's ingenious use of the length and orientation of the building, even to the extent of craftily angling the line between two pillars on the East flank of the building while maintaining its astronomical accuracy.

The line enabled Cassini not only to repeat Danti's feat of accurately dating both solstici and both equinoctia, but also to lend factual support to other scientific debates of the period. One such debate in which the Church had a primary interest was verifying the accuracy of the Gregorian calendar. While annual observation of the sun's position on the meridiana at the vernal equinox was enough to convince skeptics that Gregory was correct in eliminating ten days from the calendar in 1582, there remained the question of whether the leap year day should be eliminated in the upcoming centennial year of 1700—the first such elimination of a leap year day since the promulgation of Gregory's calendar. Cassini's measurements in 1695, however, executed with the assistance of his son Jacques and Domenico Guglielmini (1655–1710)—one of his successors at the University of Bologna—verified beyond doubt the need to eliminate the leap year day in 1700 in order to retain the astronomical accuracy of the calendar.

Figure 12. Meridian Line. Cathedral of San Petronio; Bologna, Italy.
Photo by Author. Used by permission of Basilica San Petronio.

Another of the important debates involved observations which supported the first two of Kepler's three laws of planetary motion. It had long been observed that the sun appeared to be farther away and to change more slowly in its elevation in the sky during the late spring and summer (in the Northern Hemisphere). Kepler had asserted in *Astronomia Nova* (1609) that the Earth's path around the sun was elliptical rather than round in shape, and that the sun was not in the center of the ellipse but slightly toward one side. He had further theorized that the Earth moved at irregular speeds—that it moved faster when near the sun and slower when at a greater distance. Although Cassini had to be careful in his choice of words to interpret his observations—since his work was taking place only decades after the trial of Galileo and the Copernican heliocentric view was still not officially accepted by ecclesiastical authorities—his conclusions using the meridiana at Bologna affirmed that the sun was indeed farther distant from the Earth during the northern summer, with the effect that the appearance of its elevation in our sky changed more slowly.[16]

A second meridiana was finally installed in a church in Rome in 1702. The Baths of Diocletian had long been an object of scorn by the popes and were simply left in ruin until in 1560, Pope Pius IV decided to convert the space into a church to honor the Christian martyrs who had lost their lives during Diocletian's persecution—a certain proclamation of the historical reversal of fortunes following Diocletian which had brought about the triumph of Christianity. It was none other than the famous artist Michelangelo Buonarotti (1475–1564) who was enlisted to design the Basilica Church of Santa Maria degli Angeli e dei Martyri and begin the construction, although the project extended well beyond his lifetime.

Nearly one and one-half centuries later Pope Clement XI commissioned Francesco Bianchini (1662–1729) to build a meridiana in Michelangelo's church to symbolize the victory of his predecessor Gregory's Christian calendar over the old pagan Julian calendar. It was, of course, to be longer and more useful than the small meridiana built by Danti in the Torre dei Venti more than a century previous. In addition, although every Catholic realm and nearly every Protestant realm had by this time obligingly conformed to the Gregorian dictates by eliminating the leap year day in the centennial year just past (1700), the establishment of a meridiana in Rome would give the Pope a local instrument for verifying the accuracy of the calendar. This eliminated the need for deferring to measurements taken from Cassini's meridiana at Bologna—being under the jurisdiction of an archbishop whose judgment the Pope may not always trust.

---

[16] Heilbron, 73ff; Febrizio Bònoli, "1655–2005: 350 Years of the Great Meridian Line by G. D. Cassini in the Basilica of San Petronio in Bologna," http://www.bo.astro.it/universo/cassini/meridian_ing.htm.

Figure 13. Meridian Line. Basilica Santa Maria degli Angeli, Rome. Photo by Author.

Although the gnomon at Santa Maria degli Angeli is not as high and the line not as long as Cassini's at Bologna, Bianchini's meridiana was nevertheless constructed with minute attention to detail.[17] One of those details—admittedly the original reason behind

[17] Heilbron, 152ff.

the construction of all meridiane but curiously omitted at Bologna—was the notation of the *Terminus Paschae*, or the calendrical limits of Easter. For most modern observers, it is necessary to read the line somewhat backward calendrically. Proceeding from the summer solstice—the far south end of the line nearest the wall where the gnomon admits the beam of sunlight—and counting up to day fifty-five, one observes the unmistakable papal seal of Clement XI. Day fifty-five from (before) the summer solstice is 25 April—the latest possible date of Easter. Following northward the increasing numbers on the line brings one backward calendrically to day ninety before the summer solstice—at 22 March, thirty-five days before 25 April and the earliest possible date of Easter. Hence, from the perspective of simple observation, the riddle of Easter had been finally solved.

More meridiane were constructed later in other churches, as well as in homes of nobility and in the state-sponsored observatories which were just beginning to develop during this era. Similarly, other calendrical instruments had been already developed and utilized in European cities farther north since the late fourteenth century—most notably the calendar discs attached to astronomical clocks in cathedrals in Germany and elsewhere. Consequently, what can be asserted with few qualifications is that the popular image of Christianity as a persecutor of scientists and a bastion of ignorance is in almost every case patently false. For until the rise of state-sponsored scientific establishments in several European nations in the seventeenth and eighteenth centuries Christianity in general, and the Roman Catholic Church in particular, was by far the foremost patron of scientific research. Likewise, few scientists of the era perceived a conflict between their faith and their work, but agreed with Kepler that "The plain truth is that astronomical science . . . enables the cultivated eye to behold the glory of God . . . by its practice the mind comprehends that for which God himself is praised."[18] Far from its persecutor, with few exceptions Christianity joined hand-in-hand with science as its partner and benefactor.

---

[18] Kepler, *Astronomia Nova* (Prague: P. Ramus, 1609), unnumbered introduction. Also found at http://archive.org/stream/Astronomianovaa00Kepl#page/n27/mode/1up.

Epilogue: Poisson d'Avril

Appendix G, which follows, shows the three dates for the celebration of Christ's resurrection from the year of my birth through AD 2200. Upon seeing this chart, a number of students over the years have therefore inquired as to whether the calendar date of my birth was 23 April, 6 May, or 1 April 1956. On most occasions I have resolutely adhered to the importance of my overall topic and simply stated that I was born on Easter Sunday. Only occasionally have I admitted, as some have suspected, that the calendar date of my birth was 1 April.

Most readers will readily understand my reluctance regarding such a calendar date. With the modern penchant for silly jokes and childish pranks on the first day of April, I learned at a young age that it was not wise to proclaim to everyone that it was my birthday. It was rather like having a birthday on Halloween, or Guy Fawkes's Day, or New Year's Eve, or even on April 20—which means that one's birthday is shared with Adolf Hitler. So I quickly learned the April Fool jokes and played the pranks to divert attention away from its association with my birthday, and claimed Easter Sunday as my birthday when asked. After all, what better day could one choose—a Sunday in spring when more people dress up and attend worship than at any other time of the year (with the possible exception of Christmas Eve), a day when people are prone to think about their families and what God has done for them through Christ. So Jesus and I have shared the day, and most of the time we have gotten along just fine!

Speaking of Hitler, it occurred to me one day that his birth date of 20 April put him also within the range of possibilities of having been born on Easter Sunday—since the range of dates for western Easter spans from 22 March to 25 April and the range of dates for Orthodox Easter (Gregorian Calendar) spans from 3 April to 9 May. Such a frightening prospect was somewhat relieved when I discovered that in 1889, the year of Hitler's birth, 20 April occurred on a Saturday. My sense of relief was heightened after checking the ecclesiastical calendar for 1889 and discovering that both the western Easter and the Orthodox Easter (Julian Calendar) occurred on 21 April. "Too close for comfort!" was my immediate reaction, although we might speculate how the history of the twentieth century might have been much different had Hitler realized the importance of the days surrounding his own birth.

But what about the first day of April and its association with fools? Several theories have been advanced as to the origin of this association, and this is one instance in which the modern practices are more likely the result of multiple origins rather than any single origin by itself. And what the theories of origin have in common are the varieties of folk practices connected with the beginning of spring in medieval Europe.

The theory which makes most immediate sense to those of us interested in calendars, dates, and festivals is that the modern April Fool's Day practices were begun in the aftermath of the Gregorian calendar reform in 1582, and most noticeably in France. Following the custom of many Europeans since pre-history, the vernal equinox had been celebrated as the beginning of the new year, as discussed earlier in this text. The celebrations had endured in spite of the designation of 1 January for that honor since the adoption of the Julian Calendar in 45 BC. Customarily, the new year was celebrated for an *octave*—a period of eight days—beginning with the assumed vernal equinox on 25 March. This meant that 1 April was the last day of the festival, and therefore the day when most individuals began returning home and to their usual pursuits.

But with the Gregorian calendar reform in 1582, 1 January was confirmed as the beginning of the year. Furthermore, in order to correct the calendrical drift which had slowly accrued since the creation of the Julian calendar, ten days were eliminated from October (5–14) of that year. In addition, 21 March was confirmed as the assumed vernal equinox rather than 25 March—a ruling which, as we have seen, went all the way back to the fourth century and the Council of Nicaea, but which had been conveniently ignored in the annual revelries.

The natural consequence of the calendrical reform and the gradual political reinforcement of 1 January as the official beginning of the year was that it cut the heart out of the equinoctial observances for all but scientists and theologians. No longer was there a new year to celebrate, at least not at the vernal equinox. Although in some years Easter and/or Passover came directly on the heels of the equinox, by their very nature they required a different kind of celebration. The same was true with the Christian feast of the Annunciation on 25 March. And so the equinoctial festival died a slow and sometimes painful death—a death aided, as it were, by such inventions as the *poisson d'Avril*, or fish of April.

At least intermittently during the Middle Ages, Easter had been celebrated in France as the beginning of the new year for civil and administrative purposes. With the wide variation in dates of Easter, however, it provided no end of administrative headaches for matters of taxation and political office-holding. So many welcomed the proclamation of the Gregorian calendar as a chance to begin on a clean slate with a simpler set of dates. Yet there were those who refused to give up the time-honored traditions of the equinoctial celebrations—who then became the butt of jokes such as pasting brightly-colored paper fish to their backs. They thus became known as the *poisson d'Avril*, or "fish of April," a term which has stuck to this day in France as a nickname for the first day of April.

But not everyone, even in Christianized Europe, had celebrated the beginning of a new year on 25 March. And not everyone changed in unison to the Gregorian calendar upon its pronouncement in 1582. So other explanations for the origin of April Fool's Day have been advanced which are likely as factual as the calendar change theory. Some have found a possible indication of pranks being associated with 1 April by the late fourteenth century in Chaucer's *Nun's Priest's Tale* of the vain cock Chauntecleer who is outwitted by a sly fox and nearly eaten as a result, since it is said to occur "Syn March began thritty dayes and two" ("Since March began thirty days and two"). The thirty-second day since the beginning of March would be 1 April. Others have interpreted it, however, to mean thirty-two days since March "was gone," meaning 3 May.

JESUS' BIRTH DATE AND CALENDRICAL CONSIDERATIONS IN THE TWENTY-FIRST CENTURY

Within a few generations after the adoption of Gregory's calendar, the ancient popularity of the equinoctial celebration evaporated and was forgotten. Its only remaining vestige—perhaps appropriately—was and still is the Christian feast of the Annunciation—the event discussed earlier in this text as one of the most important when considering the essential dates of Jesus' life. So what began as a pagan celebration of the divine origin of the cosmos in the unrecorded mists of ancient history was transformed and had an afterlife as a celebration of the incarnation of God into human flesh—and still focused on the Julian vernal equinox of 25 March.

In the modern world of the early twenty-first century, however, for most Christians any thoughts of the intersection of divine and human history are focused rather on the second coming of the Messiah rather than the first. This book has not attempted to engage that subject in any substantive way, although it is certainly related to any consideration of the essential dates of the life of Jesus. One need only take a glance at the plethora of books by Christian authors on bookstore shelves or initiate an online search to see the immense popularity of eschatological topics for Christians in the current generation.

As I write, we are only one year past what one well-known Christian radio broadcaster proclaimed as the sure date of Jesus' return.[1] So the very fact that you are reading this book might be sufficient evidence of the accuracy of that prediction! But this is far from the first time that such a prediction has been made. Amid the swirling debates during the late twentieth century regarding the official end of that century and the beginning of the twenty-first (and when to schedule the big bash for the new millennium), the wrangling over the alleged Y2K bug, and speculations over the social and political shape of what was to come, there was a consistent and familiar undercurrent of Christian speculation regarding the return of Jesus to set up his earthly kingdom. One popular Biblical interpretation was that Jesus' prophesied a timeline for his own return in Matthew's Gospel:

> From the fig tree learn its lesson: as soon as its branch becomes tender and puts out its leaves, you know that summer is near. So also, when you see all these things, you know that he is near, at the very gates. Truly, I say to you, this generation will not pass away until all these things take place (Matthew 24:32–34).

---

[1] "Judgment Day," http://www.familyradio.com/graphical/literature/judgment/judgment.html (discontinued).

Some Christians have interpreted the fig tree as a symbol of the nation of Israel, and "this generation" as a forty-year period beginning with the establishment of the state of Israel in 1948. Consequently, some speculated whether the return of Jesus was destined to take place by Israel's fortieth anniversary of political statehood in 1988.[2] For believers in premillennial eschatology, this meant rather conveniently that the seven-year Great Tribulation would begin by the end of 1988 (dispensationalists added that the rapture was to remove all true Christians from Earth at this time), and that Jesus would return seven years later by the end of 1995—right on cue for the approximate two-thousandth anniversary of his birth. Unfortunately, like so many other attempts to forecast the future and read the divine hand in human history, this too was proven to be nothing more than speculation.

Old ideas die hard, however, and are often re-interpreted and re-calculated instead, as evidenced by other speculations regarding possible connections between Jesus' second advent and the turn of the calendrical millennium in 2000/2001. After all, concepts such as the World Week (discussed in chapter one) have been with us a very long time, and have been enshrined in art for purposes both sacred and secular. Even with the passing of the recent millennial milestone, Christian attachment to this and a variety of other eschatological ideas is still very much alive. It is therefore not a large stretch of the imagination to wonder how many more speculations will be offered over the next two decades or more as we near the two-thousandth anniversary of Jesus' death and resurrection.[3]

---

[2] Edgar C. Whisenant, *88 Reasons Why the Rapture Could Be in 1988* (Nashville: World Bible Society, 1988). Critique of Whisenant and this entire trend in eschatological speculation are found at John Mark Ministries, "Famous Predictions of Jesus' Return," http://jmm.aaa.net.au/articles/14229.htm. Whisenant was not alone, however, as popular author Hal Lindsey also predicted the return of Jesus in the 1980s in *The Late Great Planet Earth* (Grand Rapids: Zondervan, 1998[1970]).

[3] Speculation has already become very popular among some authors concerning possible dates of Jesus' return in the first half of the twenty-first century. One of the foremost among these writers is Grant R. Jeffrey, whose early popularity with titles such as *Armageddon: Appointment with Destiny* (Colorado Springs: WaterBrook Press, 1997) has continued to the present with *The New Temple and the Second Coming: The Prophecy that Points to Christ's Return in Your Generation* (WaterBrook, 2007) and *Countdown to the Apocalypse: Learn to Read the signs that the Last Days Have Begun* (WaterBrook, 2008). Bruce Booker has analyzed this matter from the viewpoint of the Jewish feasts in "The Feasts of the Lord (Moedim)

Predictions of dates for the return of Jesus or other associated events have understandably elicited criticism, especially when stated with the certainty and finality of supposedly divine authority. A good safeguard for modern Christians, therefore, which can probably never be quoted too often, is Jesus' reminder in the verse already quoted that, "concerning that day and hour no one knows, not even the angels of heaven, nor the Son, but the Father only" (Matthew 24:36). What is often not stated as well or understood as widely, however, is the larger and more important picture—the Christian vision that the kingdoms of this world will someday "become the kingdom of our Lord and of his Christ" (Revelation 11:15). It is the Christian mandate, after all—given by Jesus himself—to teach "them to observe all that I have commanded you" (Matthew 28:20), which certainly carries with it the implication that the teachings of Jesus will have a transformational effect on human society as the numbers of his followers increase.

In other words, a follower of Christ should be marked by a passion for applying the Gospel to persons and structures in this world more than by a passion for getting themselves out of this world. This passion and vision usually do not come across in the form of a splashy announcement from a teary-eyed preacher waving a Bible above his sweaty forehead and warning of impending catastrophe, and in the modern world even less by means of military force. Historically speaking, this vision has been asserted more subtly through the establishment of Christian-based institutions involved in social welfare and community building, in education, art, and architecture, and of course through the simple sharing of the message of salvation through the death and resurrection of Christ. A plethora of such institutions is found today in thousands of locations around the world, especially in those places where large numbers of converts have heard and applied the preaching of the Gospel.

---

and their Fulfillment in Messiah Yeshua," (2008) http://www.scribd.com/doc/62705266/The-Moedim-of-Yeshua-The-Feasts-of-the-Lord-by-Bruce-Booker A thoughtful approach to Biblical statements leading to eschatological speculation is taken by Don Koenig (who himself does not agree with some of the speculations of which he writes) in "When Will Jesus Christ Return?: Projecting a Fairly Realistic Date for the Second Coming of Jesus Christ," http://www.thepropheticyears.com/comments/countdown_to_second_coming.htm.

Figure 14. Mosaic Pavement. Abbey of the Dormition, Jerusalem. Photo by Author.

One fine artistic example of the Christian vision of cultural conversion of the world is found on the floor of the sanctuary at the Abbey of the Dormition in Jerusalem. Located on Mount Zion, just outside the walls of the Old City, it is adjacent to the traditional locations of both the tomb of David and the Upper Room. Positioned above the crypt marking the location of the Virgin Mary's passing to the next life before she was later buried about a mile away at Gethsemane, the mosaic pavement on the floor of this sanctuary is a simple yet stunning statement of the Christian understanding of the intersection of human and divine history. Beginning at the center with three concentric circles and the thrice-άγιὸς (thrice-holy, a symbol of the Trinity), it expands outward as a sunburst to successively large circles featuring the names of prophets, disciples, and evangelists. The outermost circle is perhaps most telling, with its inclusion of the months of the year and the signs of the zodiac, signifying the progress of the Christian message into every aspect of human culture and understanding.

Equally significant is the calendrical orientation of the pavement, as shown in figure 15. Facing toward the front of the sanctuary and the beautiful mosaic of the Virgin Mary holding the infant Jesus which adorns the apse is none other than the month of March (Latin, "Martius"), with the accompanying zodiac symbols of Pisces and Aries. Casual modern visitors may notice that January is strangely off-center and wonder about the correctness of this orientation. It should come as no surprise, however, once we realize that it reflects once more the ancient Christian understanding of divine redemptive activity in human history. As we have observed in earlier chapters, March was understood as the first month of the year because it includes the vernal equinox—the date of the beginning of creation, and therefore of a new year. And since it was the beginning date of creation, it was also the assumed date of the incarnation, and later of the resurrection of Jesus Christ from the grave.

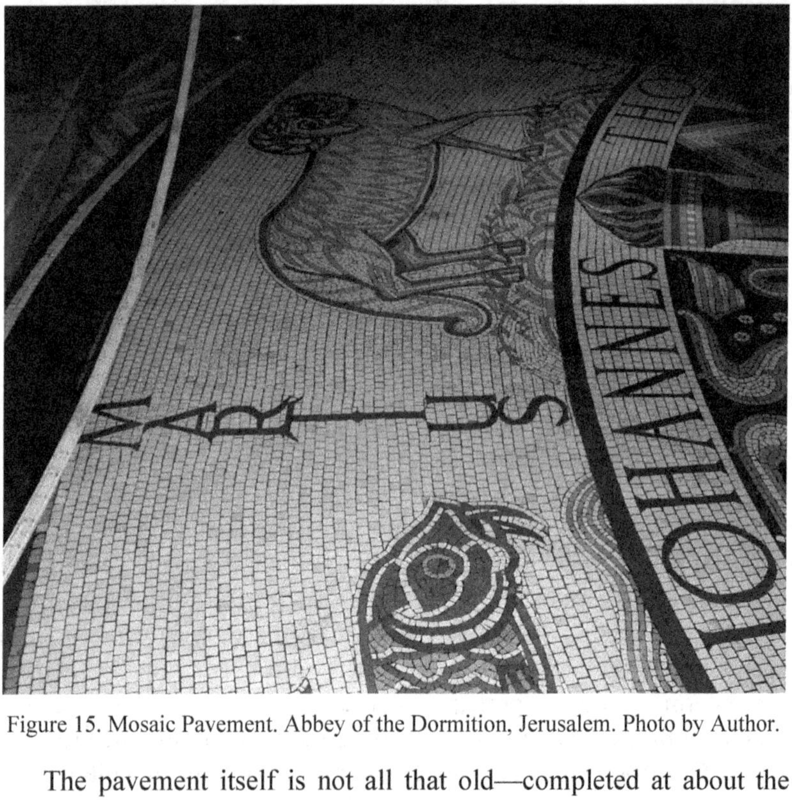

Figure 15. Mosaic Pavement. Abbey of the Dormition, Jerusalem. Photo by Author.

The pavement itself is not all that old—completed at about the beginning of the twentieth century under the direction of Father Mauritius Gisler for the German Benedictines who established an abbey on the site after it was purchased in 1898 by Kaiser Wilhelm II from Turkish sultan Abdul Hamid. Nevertheless, it is a throwback to the ancient Christian practice of utilizing the mental constructs of the surrounding culture to speak the Gospel message. While some modern Christians might balk at the use of pre-Christian and pagan symbols such as the zodiac signs, this is a reminder that early Christians were not afraid to use the understandings, signs, and symbols of the world around them as vehicles to speak the Christian message—perhaps much like Paul's use of the altar to the unknown god in Athens (Acts 17:23) as an entry point for proclaiming the Gospel to the philosophers at the Areopagus.

In addition, Christian expressions such as this are a good corrective to the modern tendency to disconnect faith from history. The Enlightenment critique of biblical and historical scholarship, which yielded positive gains and corrections in our understanding of the past, nevertheless had the unintended effect of pushing religion to the margins of human life. As a result, many modern westerners are ignorant of the centrality of religion, and especially of Christianity, in shaping the world in which they live. Even many modern Christians see their faith as a purely private matter with little relevance for the political, social, and economic world which dominates their daily existence. But this is certainly not the case when we observe the practices and public expressions of Christians of centuries ago. They had no qualms about asserting the very thing that even modern Christians purportedly celebrate every year at Christmas—the very concrete presence of God in human history. In other words, although we do not know the calendar dates of the incarnation, birth, death, and resurrection of Christ with scientific precision, we can nevertheless assert along with our ancient brothers and sisters that there was a very concrete intersection of human history with divine history. And if there was such a moment in the past when God made his presence known concretely in the midst of human history, we can rest assured that it will happen again, although at a time known to God alone.

When the research and writing for this project began several years ago, the mind of this author was fixated rather narrowly on the matter of the Paschal cycle and its historical development. As described in the preface, having been born on Easter Sunday, my initial interest in this topic was a matter of self-discovery focused on my strange but fascinating birthday. The supposition that this study could have anything more than a cursory apologetic value was rather distant from my thought processes.

But with the Christian, of course, nothing in life is just about oneself. Whether we are aware of it or not—and usually we are not—every aspect and experience of our lives has been somehow prearranged and carries a divine purpose. In most cases that purpose is known to God alone, although occasionally we are afforded small glimpses of it. Without desiring to sound like a wild-eyed visionary, it is nevertheless the conviction of this author that through the course of this project just such a glimpse of divine purpose has been afforded to him.

In other words, like many of the astronomers during the Scientific Revolution who felt that they were privileged explorers beholding divine mysteries as they gazed upon heavenly bodies never before seen by the human eye, it is the conviction of this author that there is something of the divine to behold when one views both the interdependence of the orbits of heavenly bodies and the imprecision of time measurement. It might be said that this is another example of the "hidden-ness" of God behind the details of the creation. Far from arguing for a purely materialistic conception of reality or the absence of God from the physical universe, it is this author's contention that it is only a divinely Supreme Being who could have brought such complicated, intricate, and interdependent factors into the living and workable whole that it is.

BACK TO APRIL FOOLS' DAY

Whatever the origin(s) of the modern April Fools' Day, what should be remembered is that every ancient and medieval civilization appears to have observed some form of festival during the course of the year which allowed for at least mild pranks and tomfoolery. Sometimes called "renewal festivals," they often came at the beginning of a new year or growing season, as in the case of the aforementioned Roman new year festival of Saturnalia. In the course of such festivals the social order was often temporarily overturned and mocked, with servants giving orders to masters and levying taxes upon kings and nobles. Simply stated, it was a societally acceptable way for peasants and the working classes to break out of the doldrums of winter and "blow off steam"—to feel that they were at least partially understood and appreciated by those in power. And the overall social order was not threatened, since it was understood by all to be a short-lived celebration.

For myself, the fact of having been born not only on the first day of April but also on Easter Sunday has served as an undying reminder of the impact of both the holy and the profane on the believer's life, or—to put it in Luther's terms—*simul justus et peccator*. In other words, a Christian is simultaneously *justus* (justified, or a saint) and *peccator* (a sinner). No matter how much a believer yields themselves to the lordship of Christ and the power of the Holy Spirit, he is still a sinner with all the awful possibilities which come with having a sin nature still living inside. Conversely, however, no matter how many times a believer falls into sin or fails in their spiritual duty, the grace of God won through the death and resurrection of Christ is available and sufficient for every need. And while a birthday on the first day of April is an inescapable sign to this author that he and all humans are by nature fools bent on their own folly, a birthday on Easter Sunday is an equally inescapable reminder that God loved this fool and all fools enough to provide redemption through the supreme sacrifice of his Son on our behalf.

Or, putting it in the thought forms of the ancient Christian authors consulted in the course of this project, it could be said that each of us stands at the intersection of divine history with human history. While the calendar dates of our lives may not always directly reflect events so momentous as those of the life of Christ, we are nevertheless those precious souls for whom he died and rose from the grave. Moreover, every believer is a part of the divine stream running through human history pointed toward the fulfillment of God's own purposes—purposes which are almost always hidden from human view, but of which we are sometimes afforded a precious glimpse.

Appendices

APPENDIX A

ROMAN CALENDAR CONVERTER FOR DATES OF (WESTERN) EASTER

Roman Date	***Modern (Western) Date***
11 Kalends April	22 March
10 Kalends April	23 March
9 Kalends April	24 March
8 Kalends April	25 March
7 Kalends April	26 March
6 Kalends April	27 March
5 Kalends April	28 March
4 Kalends April	29 March
3 Kalends April	30 March
Pridie (2) Kalends April	31 March
Kalends April	1 April
4 Nones April	2 April
3 Nones April	3 April
Pridie (2) Nones April	4 April
Nones April	5 April
8 Ides April	6 April
7 Ides April	7 April
6 Ides April	8 April
5 Ides April	9 April
4 Ides April	10 April
3 Ides April	11 April
Pridie (2) Ides April	12 April
Ides April	13 April
18 Kalends Mai	14 April
17 Kalends Mai	15 April
16 Kalends Mai	16 April
15 Kalends Mai	17 April
14 Kalends Mai	18 April
13 Kalends Mai	19 April
12 Kalends Mai	20 April
11 Kalends Mai	21 April
10 Kalends Mai	22 April
9 Kalends Mai	23 April
8 Kalends Mai	24 April
7 Kalends Mai	25 April

APPENDIX B

THE EASTER CYCLE OF SAINT CYRIL (AD 437–531)[1]

Year AD	*Easter Date*	*Year AD*	*Easter Date*	*Year AD*	*Easter Date*
437	11 April	469	13 April	501	22 April
438	27 March	470	5 April	502	14 April
439	16 April	471	28 March	503	30 March
440	7 April	472	16 April	504	18 April
441	23 March	473	1 April	505	10 April
442	12 April	474	21 April	506	26 March
443	4 April	475	6 April	507	15 April
444	23 April	476	28 March	508	6 April
445	8 April	477	17 April	509	22 March
446	31 March	478	9 April	510	11 April
447	20 April	479	25 March	511	3 April
448	11 April	480	13 April	512	22 April
449	27 March	481	5 April	513	7 April
450	16 April	482	25 April	514	30 March
451	8 April	483	10 April	515	19 April
452	23 March	484	1 April	516	3 April
453	12 April	485	21 April	517	26 March
454	4 April	486	6 April	518	15 April
455	24 April	487	29 March	519	31 March
456	8 April	488	17 April	520	19 April
457	31 March	489	2 April	521	11 April
458	20 April	490	25 March	522	3 April
459	5 April	491	14 April	523	16 April
460	27 March	492	5 April	524	7 April
461	16 April	493	18 April	525	30 March
462	1 April	494	10 April	526	19 April
463	21 April	495	26 March	527	4 April
464	12 April	496	14 April	528	26 March
465	28 March	497	6 April	529	15 April
466	17 April	498	29 March	530	31 March
467	9 April	499	11 April	531	20 April
468	31 March	500	2 April		

[1] These figures are Ware's estimation based upon the formula designed by Marcos J. Montes at his web site, http://www.smart.net/~mmontes/ec-cal.html. The final nineteen-year Metonic cycle included here agrees, however, with the data recorded by Dionysius Exiguus in *Liber Paschate*. See *Patrologia Latina* (J. P. Migne, ed. Montrouge: 1848), LXVII, 494. An English translation of Dionysius' Easter Cycle is found at http://hbar.phys.msu.ru/gorm/chrono/paschata.htm.

APPENDIX C

THE EASTER CYCLE OF DIONYSIUS EXIGUUS (AD 513–626)[1]

Definitions/descriptions from the columns below:

1a. Anno Diocletiani: Year since the beginning of the reign of Diocletian (AD 284).

1b. Anno Domini: Year since the assumed birth date of Jesus Christ.

Indiction: A secular Roman system of dating financial and legal documents. It had nothing to do directly with calculating Easter.

Epact: Age of the moon on 22 March, just after the vernal equinox.

Concurrent: Day of the week of 24 March, which determined the Sunday after the equinox.

Lunar/Metonic Cycle: A 19-year cycle which synchronized lunar and solar data.

Date on which the 14th paschal moon occurs: The date of Passover.

Sunday Festival: Julian date for Easter.

Age of the moon on Easter Sunday: Date in the Hebrew month of Nisan on which Easter falls.

---

[1] Much of the information for this chart was adapted from Michael Deckers, "Dionysius Exiguus. Liber de Paschate Sive Cyclus Paschalis" (http://hbar.phys.msu.ru/gorm/chrono/paschata.html), with only a few minor adjustments of details. For further information, see the explanations at the end of this chart.

Years of Diocletian	*Indiction*	*Epact*	*Concurrent*	*Lunar/ Metonic Cycle*	*Date of 14th (Paschal) Moon*	*Date of Sunday Festival (Easter)*	*Age of Moon at Easter*
229 (AD 513)	6	0	1	17	5 Apr	7 Apr	16
230 (514)	7	11	2	18	25 Mar	30 Mar	19
231 (515)	8	22	3	19	13 Apr	19 Apr	20
232 (516)	9	3	5	1	2 Apr	3 Apr	15
233 (517)	10	14	6	2	22 Mar	26 Mar	18
234 (518)	11	15	7	3	10 Apr	15 Apr	19
235 (519)	12	6	1	4	30 Mar	31 Mar	15
236 (520)	13	17	3	5	18 Apr	19 Apr	15
237 (521)	14	28	4	6	7 Apr	11 Apr	18
238 (522)	15	9	5	7	27 Mar	3 Apr	21
239 (523)	1	20	6	8	15 Apr	16 Apr	15
240 (524)	2	1	1	9	4 Apr	7 Apr	17
241 (525)	3	12	2	10	24 Mar	30 Mar	20
242 (526)	4	23	3	11	12 Apr	19 Apr	21
243 (527)	5	4	4	12	1 Apr	4 Apr	17
244 (528)	6	15	6	13	21 Mar	26 Mar	19
245 (529)	7	26	7	14	9 Apr	15 Apr	20
246 (530)	8	7	1	15	29 Mar	31 Mar	16
247 (531)	9	18	2	16	17 Apr	20 Apr	17

Years of our Lord Jesus Christ	*Indiction*	*Epact*	*Concurrent*	*Lunar/ Metonic Cycle*	*Date of 14th (Paschal) Moon*	*Date of Sunday Festival (Easter)*	*Age of Moon at Easter*
AD 532	10	0	4	17	5 Apr	11 Apr	20
533	11	11	5	18	25 Mar	27 Mar	16
534	12	22	6	19	13 Apr	16 Apr	17
535	13	3	7	1	2 Apr	8 Apr	20
536	14	14	2	2	22 Mar	23 Mar	15
537	15	25	3	3	10 Apr	12 Apr	16
538	1	6	4	4	30 Mar	4 Apr	19
539	2	17	5	5	18 Apr	24 Apr	20
540	3	28	7	6	7 Apr	8 Apr	15
541	4	9	1	7	27 Mar	31 Mar	18
542	5	20	2	8	15 Apr	20 Apr	19
543	6	1	3	9	4 Apr	5 Apr	15
544	7	12	5	10	24 Mar	27 Mar	17
545	8	23	6	11	12 Apr	16 Apr	18
546	9	4	7	12	1 Apr	8 Apr	21
547	10	15	1	13	21 Mar	24 Mar	17
548	11	26	3	14	9 Apr	12 Apr	17
549	12	7	4	15	29 Mar	4 Apr	20
550	13	18	5	16	17 Apr	24 Apr	21

AD 551	14	0	6	17	5 Apr	9 Apr	18
552	15	11	1	18	25 Mar	31 Mar	20
553	1	22	2	19	13 Apr	20 Apr	21
554	2	3	3	1	2 Apr	5 Apr	17
555	3	14	4	2	22 Mar	28 Mar	20
556	4	15	6	3	10 Apr	16 Apr	20
557	5	6	7	4	30 Mar	1 Apr	16
558	6	17	1	5	18 Apr	21 Apr	17
559	7	28	2	6	7 Apr	13 Apr	20
560	8	9	4	7	27 Mar	28 Mar	15
561	9	20	5	8	15 Apr	17 Apr	16
562	10	1	6	9	4 Apr	9 Apr	19
563	11	12	7	10	24 Mar	25 Mar	15
564	12	23	2	11	12 Apr	13 Apr	15
565	13	4	3	12	1 Apr	5 Apr	18
566	14	15	4	13	21 Mar	28 Mar	21
567	15	26	5	14	9 Apr	10 Apr	15
568	1	7	7	15	29 Mar	1 Apr	17
569	2	18	1	16	17 Apr	21 Apr	18
AD 570	3	0	2	17	5 Apr	6 Apr	15
571	4	11	3	18	25 Mar	29 Mar	18
572	5	22	5	19	13 Apr	17 Apr	18
573	6	3	6	1	2 Apr	9 Apr	21
574	7	14	7	2	22 Mar	25 Mar	17
575	8	25	1	3	10 Apr	14 Apr	18
576	9	6	3	4	30 Mar	5 Apr	20
577	10	17	4	5	18 Apr	25 Apr	21
578	11	28	5	6	7 Apr	10 Apr	17
579	12	9	6	7	27 Mar	2 Apr	20
580	13	20	1	8	15 Apr	21 Apr	20
581	14	1	2	9	4 Apr	6 Apr	16
582	15	12	3	10	24 Mar	29 Mar	19
583	1	23	4	11	12 Apr	18 Apr	20
584	2	4	6	12	1 Apr	2 Apr	15
585	3	15	7	13	21 Mar	25 Mar	18
586	4	26	1	14	9 Apr	14 Apr	19
587	5	7	2	15	29 Mar	30 Mar	15
588	6	18	4	16	17 Apr	18 Apr	15
AD 589	7	0	5	17	5 Apr	10 Apr	19
590	8	11	6	18	25 Mar	26 Mar	15
591	9	22	7	19	13 Apr	15 Apr	16
592	10	3	2	1	2 Apr	6 Apr	18
593	11	14	3	2	22 Mar	29 Mar	21
594	12	25	4	3	10 Apr	11 Apr	15
595	13	6	5	4	30 Mar	3 Apr	18
596	14	17	7	5	18 Apr	22 Apr	18
597	15	28	1	6	7 Apr	14 Apr	21

598	1	9	2	7	27 Mar	30 Mar	17
599	2	20	3	8	15 Apr	19 Apr	18
600	3	1	5	9	4 Apr	10 Apr	20
601	4	12	6	10	24 Mar	26 Mar	16
602	5	23	7	11	12 Apr	15 Apr	17
603	6	4	1	12	1 Apr	7 Apr	20
604	7	15	3	13	21 Mar	22 Mar	15
605	8	26	4	14	9 Apr	11 Apr	16
606	9	7	5	15	29 Mar	3 Apr	19
607	10	18	6	16	17 Apr	23 Apr	20
AD 608	11	0	1	17	5 Apr	7 Apr	16
609	12	11	2	18	25 Mar	30 Mar	19
610	13	22	3	19	13 Apr	19 Apr	20
611	14	3	4	1	2 Apr	4 Apr	16
612	15	14	6	2	22 Mar	26 Mar	18
613	1	25	7	3	10 Apr	15 Apr	19
614	2	6	2	4	30 Mar	31 Mar	15
615	3	17	2	5	18 Apr	20 Apr	16
616	4	28	4	6	7 Apr	11 Apr	18
617	5	9	5	7	27 Mar	3 Apr	21
618	6	20	6	8	15 Apr	16 Apr	15
619	7	1	7	9	4 Apr	8 Apr	18
620	8	12	2	10	24 Mar	30 Mar	20
621	9	28	3	11	12 Apr	19 Apr	21
622	10	4	4	12	1 Apr	4 Apr	17
623	11	15	5	13	21 Mar	27 Mar	20
624	12	26	7	14	9 Apr	15 Apr	20
625	13	7	1	15	29 Mar	31 Mar	16
626	14	18	2	16	17 Apr	20 Apr	17

APPENDIX D

VICTORIUS OF AQUITAINE VS. DIONYSIUS EXIGUUS ON THE CALCULATION OF EASTER

Victorius			Dionysius (Cyril)	
Year (AD)	*Date of Easter*	*Age of Moon*	*Date of Easter*	*Age of Moon*
513	7 April	16	7 April	16
514	30 March	19	30 March	19
515	19 April	20	19 April	20
516	3 April	15	3 April	15
517	26 March	18	26 March	18
518	15 April	19	15 April	19
519*	31 March	16	31 March	15
520*	19 April	16	19 April	15
521*	11 April	19	11 April	18
522*	3 April	22	3 April	21
523*	16 April	16	16 April	15
524*	7 April	18	7 April	17
525*	30 March	21	30 March	20
526*	19 April	22	19 April	21
527*	4 April	18	4 April	17
528*	26 March	20	26 March	19
529	15 April	21	15 April	21
530*	31 March	17	31 March	16
531*	20 April	18	20 April	17
532	11 April	20	11 April	20
533	27 March	16	27 March	16
534	16 April	17	16 April	17
535	8 April	20	8 April	20
536	23 March	15	23 March	15
537	12 April	16	12 April	16
538*	4 April	20	4 April	19
539*	24 April	21	24 April	20
540*	8 April	16	8 April	15
541*	31 March	19	31 March	18
542*	20 April	20	20 April	19
543*	5 April	16	5 April	15
544*	27 March	18	27 March	17
545*	16 April	19	16 April	18

546*	8 April	22	8 April	21
547*	24 March	18	24 March	17
548*	12 April	18	12 April	17
549*	4 April	21	4 April	20
550**	17 April	15	24 April	21
551	9 April	18	9 April	18
552	31 March	20	31 March	20
553	20 April	21	20 April	21
554	5 April	17	5 April	17
555	28 March	20	28 March	20
556	16 April	20	16 April	20
557*	1 April	17	1 April	16
558*	21 April	18	21 April	17
559*	13 April	21	13 April	20
28/560*	28 March	16	28 March	15
29/561*	17 April	17	17 April	16
30/562*	9 April	20	9 April	19
31/563*	25 March	16	25 March	15
32/564*	13 April	16	13 April	15
33/565*	5 April	19	5 April	18
34/566*	28 March	22	28 March	21
35/567*	10 April	16	10 April	15
36/568*	1 April	18	1 April	17
37/569*	21 April	19	21 April	18
38/570	6 April	15	6 April	15
39/571	29 March	18	29 March	18
40/572	17 April	18	17 April	18
41/573	9 April	21	9 April	21
42/574	25 March	17	25 March	17
43/575	14 April	18	14 April	18
44/576*	5 April	21	5 April	20
45/577**	18 April	15	25 April	21
46/578*	10 April	18	10 April	17
47/579*	2 April	21	2 April	20
48/580*	21 April	21	21 April	20
49/581*	6 April	17	6 April	16
50/582*	29 March	20	29 March	19
51/583*	18 April	21	18 April	20
52/584*	2 April	16	2 April	15
53/585*	25 March	19	25 March	18
54/586*	14 April	20	14 April	19
55/587*	30 March	16	30 March	15
56/588*	18 April	16	18 April	15
57/589	10 April	19	10 April	19
58/590	26 March	15	26 March	15
59/591	15 April	16	15 April	16
60/592	6 April	18	6 April	18

61/593	29 March	21	29 March	21
62/594	11 April	15	11 April	15
63/595*	3 April	19	3 April	18
64/596*	22 April	20	22 April	18
65/597**	7 April	15	14 April	21
66/598*	30 March	15	30 March	17
67/599*	19 April	19	19 April	18
68/600*	10 April	21	10 April	20
69/601*	26 March	17	26 March	16
70/602*	15 April	18	15 April	17
71/603*	7 April	21	7 April	20
72/604*	22 March	16	22 March	15
73/605*	11 April	17	11 April	16
74/606*	3 April	20	3 April	19
75/607*	23 April	21	23 April	20
76/608	7 April	16	7 April	16
77/609	30 March	19	30 March	19
78/610	19 April	20	19 April	20
79/611	4 April	16	4 April	16
80/612	26 March	18	26 March	18
81/613	15 April	19	15 April	19
82/614*	31 March	16	31 March	15
83/615*	20 April	17	20 April	16
84/616*	11 April	19	11 April	18
85/617**	27 March	15	3 April	21
86/618*	16 April	16	16 April	15
87/619*	8 April	19	8 April	18
88/620*	30 March	21	30 March	20
89/621**	12 April	15	19 April	21
90/622*	4 April	18	4 April	17
91/623*	27 March	21	27 March	20
92/624*	14 April	21	14 April	20
93/625*	31 March	17	31 March	16
94/626*	20 April	18	20 April	17

*Years in which the age of the moon differs.
**Years in which the date of Easter differs.

Interpretive Notes and Analyses:

See the "Roman Calendar Converter" for the Latin designations for the effective dates of (Western) Easter. All dates and figures above have been converted from the Roman calendar for the convenience of modern readers.

The dates above for Victorius of Aquitaine are taken from his *Chronica Minora* (AD 457), which included a full 532-year master cycle of dates for Easter from AD 28 (his assumed date of Jesus' death and resurrection) to 559. It is available through *Monumenta Germaniae Historica* at http://www.dmgh.de/de/fs2/object/ display/ bsb00000798_00698.html?sortIndex=010:010:0009:010:00:00. His figures for the age of the moon on the date of Easter were from measurements taken at Rome.

The dates above for Dionysius Exiguus are taken from his *Cyclus Decemnovennalis* (AD 525). It is available at http://www.hbar. phys.msu.su/gorm/chrono/paschata.htm. Dionysius' figures for the age of the moon were taken from the observatory at Alexandria, and were recognized by most of the Roman world as the more accurate.

Notice that within the relatively short 95-year span of Dionysius' Paschal Cycle (AD 532–626) Victorius' Roman figures for the age of the moon run ahead of Dionysius' Alexandrian figure rather consistently by one day, and occasionally by two. Years in which a difference in the age of the moon is recorded are marked by one asterisk [*]. Moreover, the overage in the age of the moon runs in 13-year cycles within each nineteen-year Metonic cycle: After six consecutive years of identical dates, there are thirteen consecutive years of one-day or two-day disjunction.

This might otherwise appear to be merely a small and inconsequential difference based on lunar observations at two locations, but the results are obvious. The prescribed dates for Easter do not agree five times within this ninety-five year period (AD 550, 577, 597, 617, 621). Years in which the dates for Easter disagree are marked by a double asterisk [**]. The pattern is clear: whenever the Roman figure for the age of the moon is 15 and the Alexandrian figure does not agree, the date of Easter at Rome is one week ahead of that at Alexandria. This is because the age of the moon at Easter cannot be less than 15, since Passover takes place on the fourteenth day (moon) of Nisan.

This was clearly unacceptable for Christians of the late ancient/early medieval era. Competing claims over the proper date for the celebration of the signature event of Christianity would unfortunately be tied to competing claims for jurisdiction and prominence in the church between leading cities like Rome and Alexandria. Fortunately, the Council at Nicaea in AD 325 had already recommended the use of astronomical figures from Alexandria.

It is perhaps at least partially because of Victorius' dependence on Roman rather than the ostensibly more accurate Alexandrian measurements that Dionysius was requested to construct a new Easter cycle in AD 525, although it was also to replace the nearly-expired cycle of Cyril. It is certainly plausible that Dionysius was fully aware of Victorius' work but despised it because of its dependence on faulty measurements which produced this recurring controversy. He therefore did not even mention Victorius by name, but perhaps referred to him indirectly in his prefatory letter to papal representative Petronius as one of the "aforementioned errors" which his cycle was meant to correct and supersede.

Appendix E

The Calendars of Dionysius Exiguus (AD 513–569)

The following pages show a succession of annual calendars covering the years AD 513–569. These are the years of the first three Metonic cycles included in the Paschal table of Dionysius Exiguus, as exhibited in Appendix C. To be precise, it should be stated once more that the table created by Dionysius actually covers the years AD 532–626. In order to introduce his own table, however, Dionysius included the last Metonic cycle of nineteen years from the previous Paschal table created by Cyril of Alexandria—covering the years Anno Diocletiani 229–247, which then became Anno Domini 513–531 in Dionysius' calculation of years since the birth of Jesus. I have thus included the last Metonic cycle of Cyril and the first two actually created by Dionysius.

The following calendars were created from two simple pieces of information provided by Dionysius in his Paschal table: The next-to-last column lists the annual date of Easter on the Julian calendar. The last column lists the "age of the moon," which means the date of the Jewish month of Nisan corresponding to the Julian date of Easter. Both of these calendrical details were verified by the inscriptions on the Paschal calendar stone at the Archepiscopal Museum in Ravenna.

The calendars are instrumental in that they illustrate and verify some of the interconnected factors of the calendar with the passage of time. As noted in chapter five, there has been some debate over whether Dionysius was aware of the utility of the 532-year master calendar. This question seems especially appropriate since his Easter cycle began with the year he called *Anno Domini* 532, which is effectively the first year of the second master cycle since Dionysius' purported date of Jesus' birth (1 BC). Whether he understood its utility or not, however, his figures are significant because they give us a window into the details of how the annual calendars probably looked during the lifetime of Jesus. In other words, with only slight adjustments of detail due to the passage of time, these calendars of the sixth century AD are accurate reflections of the days, months, and years of Jesus' life.

For instance, in chapter one it was established that a very plausible date for Jesus' incarnation is 30 March, 3 BC, which reflects wonderfully the Gospel message because it also happens to be the date of Passover (14 Nisan, 3758). One complete 532-year master paschal cycle later (14 Nisan, 4290) is AD 530 (shown below), where Passover correlates with 29 March. The difference of one day is explained by three calendrical factors: The first is the *saltus lunae*, or leap of the moon one day forward in every Metonic cycle. This necessitated the omission of one lunar day in every Metonic cycle (most commonly in the final year), so that in appearance the lunar occurrences slip backward one day in every Metonic cycle. This means that over the course of one master calendar cycle (28 Metonic cycles), any given lunar date such as 14 Nisan slips backward 28 days, or two days short of the entire month of Nisan. This means that, under normal conditions, 14 Nisan, 4290 would have correlated with 1 April. But this appears to complicate matters by putting the dates an additional two days apart.

Two other calendrical matters, however, explain the correlation to 29 March. As noted in chapter one, when the Romans began using the Julian calendar in 45 BC they initially made the mistake of including a leap year day (the bissextile) in every third year instead of every fourth. Hence, by 12 BC it was discovered that they had observed twelve leap years rather than the required nine, and so it was ordered by Augustus that the correction would be made by excluding the next three leap years (745, 749, and 753 AUC, or 9, 5, and 1 BC). At the likely birth date of Jesus in 3 BC, there was still a one-day overage operating in the Julian calendar. In addition, in chapter five it was explained that in spite of the four-day correction to the date of the vernal equinox made in the fourth century, the Julian calendar and its slight overage in the length of the year was still in force. This means that when Dionysius Exiguus was constructing his paschal cycle two centuries later, the Julian calendar had again advanced two days over against the Hebrew calendar. Consequently, accounting for these three days of overage in the Julian calendar means that 14 Nisan, 4290 (29 March, AD 530) is exactly one master calendar cycle past 14 Nisan 3758 (30 March, 3 BC), and therefore a very possible calendrical match for the date of Jesus' incarnation.

Likewise, it was noted in chapter five that Dionysius was very possibly attracted to AD 563 and its Easter date of 25 March as the purported 532nd anniversary of Jesus' resurrection from the dead. Unfortunately, it was also noted that although 25 March, AD 563 was linked with 15 Nisan [4323], 25 March, AD 31 was linked with 13 Nisan [3791], and that Dionysius was likely unaware that his own Julian calendar had led him astray. Such a problem evaporates, however, when one observes that the Easter date of 5 April, AD 565 corresponds to 18 Nisan, 4325. Dionysius' own table, therefore, verifies that 5 April, AD 33 (16 Nisan, 3793; the date of Jesus' resurrection from the dead) is a reasonable 532-year calendrical match for 5 April, AD 565 (18 Nisan, 4325).

A further brief word should be interjected here about the *saltus lunae*, or "leap of the moon." Since the lunar cycle accumulates one extra day over against the solar in the course of one Metonic cycle, it was the practice of computists to subtract one day from one lunation, normally in the final year of the Metonic cycle. The Venerable Bede in his *de Temporum Rationae* (42) argued that it should take place in the final lunation of that year in order to avoid confusing adjustments to any previous months. It is not known, however, how widely or consistently this practice was followed by other computists in the period before Bede.

I have nevertheless followed Bede's rule in applying the saltus lunae to the following calendars in two of the three Metonic cycles covered. The reader will notice that in AD 550 and 569—both of which were final years in the Metonic cycles of Dionysius' Paschal table—the Hebrew month of Kislev has only 28 days rather than its usual 29 or 30 days. The rule cannot be followed rigidly, however, as is illustrated in the first Metonic cycle of Dionysius' table (AD 513–531). In this case, the saltus is included not in the final year of the cycle but eleven years earlier in AD 520. This is due to the fact that AD 532—the first year of the next Metonic cycle—was a leap year, thereby preventing the saltus in the year previous.

These calendars are likewise useful in that they allow the tracking of other details as it regards the conjunction of the Julian and Hebrew calendars for this period of time. One such example is the annual Julian date of Yom Kippur. Similar to the dates of Passover and Easter, Yom Kippur appears to wander significantly through September and October. Whatever date one wishes to track, this offers one small example of how the Jewish community of this era likely inserted the extra lunar month seven times in every nineteen years, and how that insertion, or the lack thereof, affected the flow of Hebrew dates over against the Julian calendar. If nothing else, this gives us a fairly accurate calendrical view of the very days of the life of Jesus.

AD 513/Dionysius' Cycle 0, Year 1/Metonic Year 17/Easter 07 April, Luna 16

S	*M*	*T*	*W*	*T*	*F*	*S*		*S*	*M*	*T*	*W*	*T*	*F*	*S*
		1	2	3	4	5	Tebeth>			9	10	11	12	13
6	7	8	9	10	11	12	<Jan	14	15	16	17	18	19	20
13	14	15	16	17	18	19		21	22	23	24	25	26	27
20	21	22	23	24	25	26	Shevat>	28	29	1	2	3	4	5
27	28	29	30	31	1	2		6	7	8	9	10	11	12
3	4	5	6	7	8	9	<Feb	13	14	15	16	17	18	19
10	11	12	13	14	15	16		20	21	22	23	24	25	26
17	18	19	20	21	22	23		27	28	29	30	1	2	3
24	25	26	27	28	1	2	Adar>	4	5	6	7	8	9	10
3	4	5	6	7	8	9	<March	11	12	13	14	15	16	17
10	11	12	13	14	15	16		18	19	20	21	22	23	24
17	18	19	20	21	22	23		25	26	27	28	29	30	1
24	25	26	27	28	29	30	Nisan>	2	3	4	5	6	7	8
31	1	2	3	4	5	6	<April	9	10	11	12	13	14	15
7	8	9	10	11	12	13		16	17	18	19	20	21	22
14	15	16	17	18	19	20		23	24	25	26	27	28	29
21	22	23	24	25	26	27	Iyar>	30	1	2	3	4	5	6
28	29	30	1	2	3	4	<May	7	8	9	10	11	12	13
5	6	7	8	9	10	11		14	15	16	17	18	19	20
12	13	14	15	16	17	18		21	22	23	24	25	26	27
19	20	21	22	23	24	25	Sivan>	28	29	1	2	3	4	5
26	27	28	29	30	31	1		6	7	8	9	10	11	12
2	3	4	5	6	7	8	<June	13	14	15	16	17	18	19
9	10	11	12	13	14	15		20	21	22	23	24	25	26
16	17	18	19	20	21	22		27	28	29	30	1	2	3
23	24	25	26	27	28	29	Tammuz>	4	5	6	7	8	9	10
30	1	2	3	4	5	6	<July	11	12	13	14	15	16	17
7	8	9	10	11	12	13		18	19	20	21	22	23	24
14	15	16	17	18	19	20		25	26	27	28	29	1	2
21	22	23	24	25	26	27	Ab>	3	4	5	6	7	8	9
28	29	30	31	1	2	3	<August	10	11	12	13	14	15	16
4	5	6	7	8	9	10		17	18	19	20	21	22	23
11	12	13	14	15	16	17		24	25	26	27	28	29	30
18	19	20	21	22	23	24	Elul>	1	2	3	4	5	6	7
25	26	27	28	29	30	31		8	9	10	11	12	13	14
1	2	3	4	5	6	7	<Sept	15	16	17	18	19	20	21
8	9	10	11	12	13	14		22	23	24	25	26	27	28
15	16	17	18	19	20	21	Tishri>	29	1	2	3	4	5	6
22	23	24	25	26	27	28		7	8	9	10	11	12	13
29	30	1	2	3	4	5	<Oct	14	15	16	17	18	19	20
6	7	8	9	10	11	12		21	22	23	24	25	26	27
13	14	15	16	17	18	19		28	29	30	1	2	3	4
20	21	22	23	24	25	26	Heshvan>	5	6	7	8	9	10	11
27	28	29	30	31	1	2		12	13	14	15	16	17	18
3	4	5	6	7	8	9	<Nov	19	20	21	22	23	24	25
10	11	12	13	14	15	16		26	27	28	29	1	2	3
17	18	19	20	21	22	23	Kislev>	4	5	6	7	8	9	10
24	25	26	27	28	29	30		11	12	13	14	15	16	17
1	2	3	4	5	6	7	<Dec	18	19	20	21	22	23	24
8	9	10	11	12	13	14		25	26	27	28	29	1	2
15	16	17	18	19	20	21	Tebeth>	3	4	5	6	7	8	9
22	23	24	25	26	27	28		10	11	12	13	14	15	16
29	30	31						17	18	19				

AD 514/Dionysius' Cycle 0, Year 2/Metonic Year 18/Easter 30 March, Luna 19

S	*M*	*T*	*W*	*T*	*F*	*S*		*S*	*M*	*T*	*W*	*T*	*F*	*S*
			1	2	3	4	Tebeth>				20	21	22	23
5	6	7	8	9	10	11	<Jan	24	25	26	27	28	29	1
12	13	14	15	16	17	18	Shevat>	2	3	4	5	6	7	8
19	20	21	22	23	24	25		9	10	11	12	13	14	15
26	27	28	29	30	31	1		16	17	18	19	20	21	22
2	3	4	5	6	7	8	<Feb	23	24	25	26	27	28	29
9	10	11	12	13	14	15	Adar>	30	1	2	3	4	5	6
16	17	18	19	20	21	22		7	8	9	10	11	12	13
23	24	25	26	27	28	1		14	15	16	17	18	19	20
2	3	4	5	6	7	8	<March	21	22	23	24	25	26	27
9	10	11	12	13	14	15	Nisan>	28	29	30	1	2	3	4
16	17	18	19	20	21	22		5	6	7	8	9	10	11
23	24	25	26	27	28	29		12	13	14	15	16	17	18
30	31	1	2	3	4	5	<April	19	20	21	22	23	24	25
6	7	8	9	10	11	12		26	27	28	29	30	1	2
13	14	15	16	17	18	19	Iyar>	3	4	5	6	7	8	9
20	21	22	23	24	25	26		10	11	12	13	14	15	16
27	28	29	30	1	2	3		17	18	19	20	21	22	23
4	5	6	7	8	9	10	<May	24	25	26	27	28	29	1
11	12	13	14	15	16	17	Sivan>	2	3	4	5	6	7	8
18	19	20	21	22	23	24		9	10	11	12	13	14	15
25	26	27	28	29	30	31		16	17	18	19	20	21	22
1	2	3	4	5	6	7	<June	23	24	25	26	27	28	29
8	9	10	11	12	13	14	Tammuz>	30	1	2	3	4	5	6
15	16	17	18	19	20	21		7	8	9	10	11	12	13
22	23	24	25	26	27	28		14	15	16	17	18	19	20
29	30	1	2	3	4	5	<July	21	22	23	24	25	26	27
6	7	8	9	10	11	12		28	29	1	2	3	4	5
13	14	15	16	17	18	19	Ab>	6	7	8	9	10	11	12
20	21	22	23	24	25	26		13	14	15	16	17	18	19
27	28	29	30	31	1	2		20	21	22	23	24	25	26
3	4	5	6	7	8	9	<August	27	28	29	30	1	2	3
10	11	12	13	14	15	16	Elul>	4	5	6	7	8	9	10
17	18	19	20	21	22	23		11	12	13	14	15	16	17
24	25	26	27	28	29	30		18	19	20	21	22	23	24
31	1	2	3	4	5	6	<Sept	25	26	27	28	29	1	2
7	8	9	10	11	12	13	Tishri>	3	4	5	6	7	8	9
14	15	16	17	18	19	20		10	11	12	13	14	15	16
21	22	23	24	25	26	27		17	18	19	20	21	22	23
28	29	30	1	2	3	4		24	25	26	27	28	29	30
5	6	7	8	9	10	11	<Oct	1	2	3	4	5	6	7
12	13	14	15	16	17	18	Heshvan>	8	9	10	11	12	13	14
19	20	21	22	23	24	25		15	16	17	18	19	20	21
26	27	28	29	30	31	1		22	23	24	25	26	27	28
2	3	4	5	6	7	8	<Nov	29	1	2	3	4	5	6
9	10	11	12	13	14	15	Kislev>	7	8	9	10	11	12	13
16	17	18	19	20	21	22		14	15	16	17	18	19	20
23	24	25	26	27	28	29		21	22	23	24	25	26	27
30	1	2	3	4	5	6	<Dec	28	29	30	1	2	3	4
7	8	9	10	11	12	13	Tebeth>	5	6	7	8	9	10	11
14	15	16	17	18	19	20		12	13	14	15	16	17	18
21	22	23	24	25	26	27		19	20	21	22	23	24	25
28	29	30	31					26	27	28	29			

AD 515/Dionysius' Cycle 0, Year 3 (emb.)/Metonic Year 19/Easter 19 April, Luna 20

S	*M*	*T*	*W*	*T*	*F*	*S*		*S*	*M*	*T*	*W*	*T*	*F*	*S*
				1	2	3	Shevat>				S	1	2	3
4	5	6	7	8	9	10	<Jan	4	5	6	7	8	9	10
11	12	13	14	15	16	17		11	12	13	14	15	16	17
18	19	20	21	22	23	24		18	19	20	21	22	23	24
25	26	27	28	29	30	31	Adar>	25	26	27	28	29	30	1
1	2	3	4	5	6	7	<Feb	2	3	4	5	6	7	8
8	9	10	11	12	13	14		9	10	11	12	13	14	15
15	16	17	18	19	20	21		16	17	18	19	20	21	22
22	23	24	25	26	27	28		23	24	25	26	27	28	29
1	2	3	4	5	6	7	<March	30	1	2	3	4	5	6
8	9	10	11	12	13	14	Adar 2>	7	8	9	10	11	12	13
15	16	17	18	19	20	21		14	15	16	17	18	19	20
22	23	24	25	26	27	28		21	22	23	24	25	26	27
29	30	31	1	2	3	4		28	29	1	2	3	4	5
5	6	7	8	9	10	11	<April	6	7	8	9	10	11	12
12	13	14	15	16	17	18		13	14	15	16	17	18	19
19	20	21	22	23	24	25		20	21	22	23	24	25	26
26	27	28	29	30	1	2	Iyar>	27	28	29	30	1	2	3
3	4	5	6	7	8	9	<May	4	5	6	7	8	9	10
10	11	12	13	14	15	16		11	12	13	14	15	16	17
17	18	19	20	21	22	23		18	19	20	21	22	23	24
24	25	26	27	28	29	30	Sivan>	25	26	27	28	29	1	2
31	1	2	3	4	5	6	<June	3	4	5	6	7	8	9
7	8	9	10	11	12	13		10	11	12	13	14	15	16
14	15	16	17	18	19	20		17	18	19	20	21	22	23
21	22	23	24	25	26	27		24	25	26	27	28	29	30
28	29	30	1	2	3	4	Tammuz>	1	2	3	4	5	6	7
5	6	7	8	9	10	11	<July	8	9	10	11	12	13	14
12	13	14	15	16	17	18		15	16	17	18	19	20	21
19	20	21	22	23	24	25		22	23	24	25	26	27	28
26	27	28	29	30	31	1	Ab>	29	1	2	3	4	5	6
2	3	4	5	6	7	8	<August	7	8	9	10	11	12	13
9	10	11	12	13	14	15		14	15	16	17	18	19	20
16	17	18	19	20	21	22		21	22	23	24	25	26	27
23	24	25	26	27	28	29	Elul>	28	29	30	1	2	3	4
30	31	1	2	3	4	5	<Sept	5	6	7	8	9	10	11
6	7	8	9	10	11	12		12	13	14	15	16	17	18
13	14	15	16	17	18	19		19	20	21	22	23	24	25
20	21	22	23	24	25	26		26	27	28	29	1	2	3
27	28	29	30	1	2	3	Tishri>	4	5	6	7	8	9	10
4	5	6	7	8	9	10	<Oct	11	12	13	14	15	16	17
11	12	13	14	15	16	17		18	19	20	21	22	23	24
18	19	20	21	22	23	24		25	26	27	28	29	30	1
25	26	27	28	29	30	1	Heshvan>	2	3	4	5	6	7	8
2	3	4	5	6	7	8	<Nov	9	10	11	12	13	14	15
9	10	11	12	13	14	15		16	17	18	19	20	21	22
16	17	18	19	20	21	22		23	24	25	26	27	28	29
23	24	25	26	27	28	29	Kislev>	1	2	3	4	5	6	7
30	1	2	3	4	5	6	<Dec	8	9	10	11	12	13	14
7	8	9	10	11	12	13		15	16	17	18	19	20	21
14	15	16	17	18	19	20		22	23	24	25	26	27	28
21	22	23	24	25	26	27	Tebeth>	29	1	2	3	4	5	6
28	29	30	31					7	8	9	10			

AD 516/Dionysius' Cycle 0, Year 4/Metonic Year 1/Easter 03 April, Luna 15

S	*M*	*T*	*W*	*T*	*F*	*S*		*S*	*M*	*T*	*W*	*T*	*F*	*S*
					1	2	Tebeth>						11	12
3	4	5	6	7	8	9	<Jan	13	14	15	16	17	18	19
10	11	12	13	14	15	16		20	21	22	23	24	25	26
17	18	19	20	21	22	23		27	28	29	1	2	3	4
24	25	26	27	28	29	30	Shevat>	5	6	7	8	9	10	11
31	1	2	3	4	5	6	<Feb	12	13	14	15	16	17	18
7	8	9	10	11	12	13		19	20	21	22	23	24	25
14	15	16	17	18	19	20		26	27	28	29	30	1	2
21	22	23	24	25	26	27	Adar>	3	4	5	6	7	8	9
28	29	1	2	3	4	5	<March	10	11	12	13	14	15	16
6	7	8	9	10	11	12		17	18	19	20	21	22	23
13	14	15	16	17	18	19		24	25	26	27	28	29	30
20	21	22	23	24	25	26	Nisan>	1	2	3	4	5	6	7
27	28	29	30	31	1	2		8	9	10	11	12	13	14
3	4	5	6	7	8	9	<April	15	16	17	18	19	20	21
10	11	12	13	14	15	16		22	23	24	25	26	27	28
17	18	19	20	21	22	23	Iyar>	29	30	1	2	3	4	5
24	25	26	27	28	29	30		6	7	8	9	10	11	12
1	2	3	4	5	6	7	<May	13	14	15	16	17	18	19
8	9	10	11	12	13	14		20	21	22	23	24	25	26
15	16	17	18	19	20	21		27	28	29	1	2	3	4
22	23	24	25	26	27	28	Sivan>	5	6	7	8	9	10	11
29	30	31	1	2	3	4	<June	12	13	14	15	16	17	18
5	6	7	8	9	10	11		19	20	21	22	23	24	25
12	13	14	15	16	17	18		26	27	28	29	30	1	2
19	20	21	22	23	24	25	Tammuz>	3	4	5	6	7	8	9
26	27	28	29	30	1	2		10	11	12	13	14	15	16
3	4	5	6	7	8	9	<July	17	18	19	20	21	22	23
10	11	12	13	14	15	16		24	25	26	27	28	29	1
17	18	19	20	21	22	23	Ab>	2	3	4	5	6	7	8
24	25	26	27	28	29	30		9	10	11	12	13	14	15
31	1	2	3	4	5	6	<August	16	17	18	19	20	21	22
7	8	9	10	11	12	13		23	24	25	26	27	28	29
14	15	16	17	18	19	20	Elul>	30	1	2	3	4	5	6
21	22	23	24	25	26	27		7	8	9	10	11	12	13
28	29	30	31	1	2	3		14	15	16	17	18	19	20
4	5	6	7	8	9	10	<Sept	21	22	23	24	25	26	27
11	12	13	14	15	16	17		28	29	1	2	3	4	5
18	19	20	21	22	23	24	Tishri>	6	7	8	9	10	11	12
25	26	27	28	29	30	1		13	14	15	16	17	18	19
2	3	4	5	6	7	8	<Oct	20	21	22	23	24	25	26
9	10	11	12	13	14	15		27	28	29	30	1	2	3
16	17	18	19	20	21	22	Heshvan>	4	5	6	7	8	9	10
23	24	25	26	27	28	29		11	12	13	14	15	16	17
30	31	1	2	3	4	5	<Nov	18	19	20	21	22	23	24
6	7	8	9	10	11	12		25	26	27	28	29	1	2
13	14	15	16	17	18	19	Kislev>	3	4	5	6	7	8	9
20	21	22	23	24	25	26		10	11	12	13	14	15	16
27	28	29	30	1	2	3		17	18	19	20	21	22	23
4	5	6	7	8	9	10	<Dec	24	25	26	27	28	29	1
11	12	13	14	15	16	17	Tebeth>	2	3	4	5	6	7	8
18	19	20	21	22	23	24		9	10	11	12	13	14	15
25	26	27	28	29	30	31		16	17	18	19	20	21	22

AD 517/Dionysius' Cycle 0, Year 5/Metonic Year 2/Easter 26 March, Luna 18

S	*M*	*T*	*W*	*T*	*F*	*S*		*S*	*M*	*T*	*W*	*T*	*F*	*S*
1	2	3	4	5	6	7	Tebeth>	23	24	25	26	27	28	29
8	9	10	11	12	13	14	<Jan	1	2	3	4	5	6	7
15	16	17	18	19	20	21	Shevat>	8	9	10	11	12	13	14
22	23	24	25	26	27	28		15	16	17	18	19	20	21
29	30	31	1	2	3	4		22	23	24	25	26	27	28
5	6	7	8	9	10	11	<Feb	29	30	1	2	3	4	5
12	13	14	15	16	17	18	Adar>	6	7	8	9	10	11	12
19	20	21	22	23	24	25		13	14	15	16	17	18	19
26	27	28	1	2	3	4		20	21	22	23	24	25	26
5	6	7	8	9	10	11	<March	27	28	29	30	1	2	3
12	13	14	15	16	17	18	Nisan>	4	5	6	7	8	9	10
19	20	21	22	23	24	25		11	12	13	14	15	16	17
26	27	28	29	30	31	1		18	19	20	21	22	23	24
2	3	4	5	6	7	8	<April	25	26	27	28	29	30	1
9	10	11	12	13	14	15	Iyar>	2	3	4	5	6	7	8
16	17	18	19	20	21	22		9	10	11	12	13	14	15
23	24	25	26	27	28	29		16	17	18	19	20	21	22
30	1	2	3	4	5	6	<May	23	24	25	26	27	28	29
7	8	9	10	11	12	13	Sivan>	1	2	3	4	5	6	7
14	15	16	17	18	19	20		8	9	10	11	12	13	14
21	22	23	24	25	26	27		15	16	17	18	19	20	21
28	29	30	31	1	2	3		22	23	24	25	26	27	28
4	5	6	7	8	9	10	<June	29	30	1	2	3	4	5
11	12	13	14	15	16	17	Tammuz>	6	7	8	9	10	11	12
18	19	20	21	22	23	24		13	14	15	16	17	18	19
25	26	27	28	29	30	1		20	21	22	23	24	25	26
2	3	4	5	6	7	8	<July	27	28	29	1	2	3	4
9	10	11	12	13	14	15	Ab>	5	6	7	8	9	10	11
16	17	18	19	20	21	22		12	13	14	15	16	17	18
23	24	25	26	27	28	29		19	20	21	22	23	24	25
30	31	1	2	3	4	5	<August	26	27	28	29	30	1	2
6	7	8	9	10	11	12	Elul>	3	4	5	6	7	8	9
13	14	15	16	17	18	19		10	11	12	13	14	15	16
20	21	22	23	24	25	26		17	18	19	20	21	22	23
27	28	29	30	31	1	2	<Sept	24	25	26	27	28	29	1
3	4	5	6	7	8	9	Tishri>	2	3	4	5	6	7	8
10	11	12	13	14	15	16		9	10	11	12	13	14	15
17	18	19	20	21	22	23		16	17	18	19	20	21	22
24	25	26	27	28	29	30		23	24	25	26	27	28	29
1	2	3	4	5	6	7	<Oct	30	1	2	3	4	5	6
8	9	10	11	12	13	14	Heshvan>	7	8	9	10	11	12	13
15	16	17	18	19	20	21		14	15	16	17	18	19	20
22	23	24	25	26	27	28		21	22	23	24	25	26	27
29	30	31	1	2	3	4	<Nov	28	29	1	2	3	4	5
5	6	7	8	9	10	11	Kislev>	6	7	8	9	10	11	12
12	13	14	15	16	17	18		13	14	15	16	17	18	19
19	20	21	22	23	24	25		20	21	22	23	24	25	26
26	27	28	29	30	1	2	Tebeth>	27	28	29	30	1	2	3
3	4	5	6	7	8	9	<Dec	4	5	6	7	8	9	10
10	11	12	13	14	15	16		11	12	13	14	15	16	17
17	18	19	20	21	22	23		18	19	20	21	22	23	24
24	25	26	27	28	29	30		25	26	27	28	29	1	2
31							Shevat>	3						

AD 518/Dionysius' Cycle 0, Year 6 (emb.)/Metonic Year 3/Easter 15 April, Luna 19

S	*M*	*T*	*W*	*T*	*F*	*S*		*S*	*M*	*T*	*W*	*T*	*F*	*S*
	1	2	3	4	5	6	Shevat>		4	5	6	7	8	9
7	8	9	10	11	12	13	<Jan	10	11	12	13	14	15	16
14	15	16	17	18	19	20		17	18	19	20	21	22	23
21	22	23	24	25	26	27		24	25	26	27	28	29	30
28	29	30	31	1	2	3	Adar>	1	2	3	4	5	6	7
4	5	6	7	8	9	10	<Feb	8	9	10	11	12	13	14
11	12	13	14	15	16	17		15	16	17	18	19	20	21
18	19	20	21	22	23	24		22	23	24	25	26	27	28
25	26	27	28	1	2	3	Adar 2>	29	30	1	2	3	4	5
4	5	6	7	8	9	10	<March	6	7	8	9	10	11	12
11	12	13	14	15	16	17		13	14	15	16	17	18	19
18	19	20	21	22	23	24		20	21	22	23	24	25	26
25	26	27	28	29	30	31	Nisan>	27	28	29	1	2	3	4
1	2	3	4	5	6	7	<April	5	6	7	8	9	10	11
8	9	10	11	12	13	14		12	13	14	15	16	17	18
15	16	17	18	19	20	21		19	20	21	22	23	24	25
22	23	24	25	26	27	28		26	27	28	29	30	1	2
29	30	1	2	3	4	5	Iyar>	3	4	5	6	7	8	9
6	7	8	9	10	11	12	<May	10	11	12	13	14	15	16
13	14	15	16	17	18	19		17	18	19	20	21	22	23
20	21	22	23	24	25	26		24	25	26	27	28	29	1
27	28	29	30	31	1	2	Sivan>	2	3	4	5	6	7	8
3	4	5	6	7	8	9	<June	9	10	11	12	13	14	15
10	11	12	13	14	15	16		16	17	18	19	20	21	22
17	18	19	20	21	22	23		23	24	25	26	27	28	29
24	25	26	27	28	29	30	Tammuz>	30	1	2	3	4	5	6
1	2	3	4	5	6	7	<July	7	8	9	10	11	12	13
8	9	10	11	12	13	14		14	15	16	17	18	19	20
15	16	17	18	19	20	21		21	22	23	24	25	26	27
22	23	24	25	26	27	28	Ab>	28	29	1	2	3	4	5
29	30	31	1	2	3	4	<August	6	7	8	9	10	11	12
5	6	7	8	9	10	11		13	14	15	16	17	18	19
12	13	14	15	16	17	18		20	21	22	23	24	25	26
19	20	21	22	23	24	25		27	28	29	30	1	2	3
26	27	28	29	30	31	1	Elul>	4	5	6	7	8	9	10
2	3	4	5	6	7	8	<Sept	11	12	13	14	15	16	17
9	10	11	12	13	14	15		18	19	20	21	22	23	24
16	17	18	19	20	21	22		25	26	27	28	29	1	2
23	24	25	26	27	28	29	Tishri>	3	4	5	6	7	8	9
30	1	2	3	4	5	6	<Oct	10	11	12	13	14	15	16
7	8	9	10	11	12	13		17	18	19	20	21	22	23
14	15	16	17	18	19	20		24	25	26	27	28	29	30
21	22	23	24	25	26	27	Heshvan>	1	2	3	4	5	6	7
28	29	30	31	1	2	3		8	9	10	11	12	13	14
4	5	6	7	8	9	10	<Nov	15	16	17	18	19	20	21
11	12	13	14	15	16	17		22	23	24	25	26	27	28
18	19	20	21	22	23	24	Kislev>	29	1	2	3	4	5	6
25	26	27	28	29	30	1		7	8	9	10	11	12	13
2	3	4	5	6	7	8	<Dec	14	15	16	17	18	19	20
9	10	11	12	13	14	15		21	22	23	24	25	26	27
16	17	18	19	20	21	22		28	29	1	2	3	4	5
23	24	25	26	27	28	29	Tebeth>	6	7	8	9	10	11	12
30	31							13	14					

AD 519/Dionysius' Cycle 0, Year 7/Metonic Year 4/Easter 31 March, Luna 15

S	*M*	*T*	*W*	*T*	*F*	*S*		*S*	*M*	*T*	*W*	*T*	*F*	*S*
		1	2	3	4	5	Tebeth>			15	16	17	18	19
6	7	8	9	10	11	12	<Jan	20	21	22	23	24	25	26
13	14	15	16	17	18	19		27	28	29	1	2	3	4
20	21	22	23	24	25	26	Shevat>	5	6	7	8	9	10	11
27	28	29	30	31	1	2		12	13	14	15	16	17	18
3	4	5	6	7	8	9	<Feb	19	20	21	22	23	24	25
10	11	12	13	14	15	16		26	27	28	29	30	1	2
17	18	19	20	21	22	23	Adar>	3	4	5	6	7	8	9
24	25	26	27	28	1	2		10	11	12	13	14	15	16
3	4	5	6	7	8	9	<Mar	17	18	19	20	21	22	23
10	11	12	13	14	15	16		24	25	26	27	28	29	30
17	18	19	20	21	22	23	Nisan>	1	2	3	4	5	6	7
24	25	26	27	28	29	30		8	9	10	11	12	13	14
31	1	2	3	4	5	6	<April	15	16	17	18	19	20	21
7	8	9	10	11	12	13		22	23	24	25	26	27	28
14	15	16	17	18	19	20	Iyar>	29	30	1	2	3	4	5
21	22	23	24	25	26	27		6	7	8	9	10	11	12
28	29	30	1	2	3	4	<May	13	14	15	16	17	18	19
5	6	7	8	9	10	11		20	21	22	23	24	25	26
12	13	14	15	16	17	18		27	28	29	1	2	3	4
19	20	21	22	23	24	25	Sivan>	5	6	7	8	9	10	11
26	27	28	29	30	31	1		12	13	14	15	16	17	18
2	3	4	5	6	7	8	<June	19	20	21	22	23	24	25
9	10	11	12	13	14	15		26	27	28	29	30	1	2
16	17	18	19	20	21	22	Tammuz>	3	4	5	6	7	8	9
23	24	25	26	27	28	29		10	11	12	13	14	15	16
30	1	2	3	4	5	6	<July	17	18	19	20	21	22	23
7	8	9	10	11	12	13		24	25	26	27	28	29	1
14	15	16	17	18	19	20	Ab>	2	3	4	5	6	7	8
21	22	23	24	25	26	27		9	10	11	12	13	14	15
28	29	30	31	1	2	3		16	17	18	19	20	21	22
4	5	6	7	8	9	10	<August	23	24	25	26	27	28	29
11	12	13	14	15	16	17	Elul>	30	1	2	3	4	5	6
18	19	20	21	22	23	24		7	8	9	10	11	12	13
25	26	27	28	29	30	31		14	15	16	17	18	19	20
1	2	3	4	5	6	7	<Sept	21	22	23	24	25	26	27
8	9	10	11	12	13	14		27	28	29	1	2	3	4
15	16	17	18	19	20	21	Tishri>	5	6	7	8	9	10	11
22	23	24	25	26	27	28		12	13	14	15	16	17	18
29	30	1	2	3	4	5	<Oct	19	20	21	22	23	24	25
6	7	8	9	10	11	12		26	27	28	29	30	1	2
13	14	15	16	17	18	19	Heshvan>	3	4	5	6	7	8	9
20	21	22	23	24	25	26		10	11	12	13	14	15	16
27	28	29	30	31	1	2		17	18	19	20	21	22	23
3	4	5	6	7	8	9	<Nov	24	25	26	27	28	29	1
10	11	12	13	14	15	16	Kislev>	2	3	4	5	6	7	8
17	18	19	20	21	22	23		9	10	11	12	13	14	15
24	25	26	27	28	29	30		16	17	18	19	20	21	22
1	2	3	4	5	6	7	<Dec	23	24	25	26	27	28	29
8	9	10	11	12	13	14	Tebeth>	30	1	2	3	4	5	6
15	16	17	18	19	20	21		7	8	9	10	11	12	13
22	23	24	25	26	27	28		14	15	16	17	18	19	20
29	30	31						21	22	23				

AD 520/Dionysius' Cycle 0, Year 8 (emb.)/Metonic Year 5/Easter 19 April, Luna 15

S	*M*	*T*	*W*	*T*	*F*	*S*		*S*	*M*	*T*	*W*	*T*	*F*	*S*
			1	2	3	4	Tebeth>				24	25	26	27
5	6	7	8	9	10	11	<Jan	28	29	1	2	3	4	5
12	13	14	15	16	17	18	Shevat>	6	7	8	9	10	11	12
19	20	21	22	23	24	25		13	14	15	16	17	18	19
26	27	28	29	30	31	1		20	21	22	23	24	25	26
2	3	4	5	6	7	8	<Feb	27	28	29	30	1	2	3
9	10	11	12	13	14	15	Adar>	4	5	6	7	8	9	10
16	17	18	19	20	21	22		11	12	13	14	15	16	17
23	24	25	26	27	28	29		18	19	20	21	22	23	24
1	2	3	4	5	6	7	<Mar	25	26	27	28	29	30	1
8	9	10	11	12	13	14	Adar 2>	2	3	4	5	6	7	8
15	16	17	18	19	20	21		9	10	11	12	13	14	15
22	23	24	25	26	27	28		16	17	18	19	20	21	22
29	30	31	1	2	3	4	<April	23	24	25	26	27	28	29
5	6	7	8	9	10	11	Nisan>	1	2	3	4	5	6	7
12	13	14	15	16	17	18		8	9	10	11	12	13	14
19	20	21	22	23	24	25		15	16	17	18	19	20	21
26	27	28	29	30	1	2		22	23	24	25	26	27	28
3	4	5	6	7	8	9	<May	29	30	1	2	3	4	5
10	11	12	13	14	15	16	Iyar>	6	7	8	9	10	11	12
17	18	19	20	21	22	23		13	14	15	16	17	18	19
24	25	26	27	28	29	30		20	21	22	23	24	25	26
31	1	2	3	4	5	6	<June	27	28	29	1	2	3	4
7	8	9	10	11	12	13	Sivan>	5	6	7	8	9	10	11
14	15	16	17	18	19	20		12	13	14	15	16	17	18
21	22	23	24	25	26	27		19	20	21	22	23	24	25
28	29	30	1	2	3	4	<July	26	27	28	29	30	1	2
5	6	7	8	9	10	11	Tammuz>	3	4	5	6	7	8	9
12	13	14	15	16	17	18		10	11	12	13	14	15	16
19	20	21	22	23	24	25		17	18	19	20	21	22	23
26	27	28	29	30	31	1		24	25	26	27	28	29	1
2	3	4	5	6	7	8	<August	2	3	4	5	6	7	8
9	10	11	12	13	14	15	Ab>	9	10	11	12	13	14	15
16	17	18	19	20	21	22		16	17	18	19	20	21	22
23	24	25	26	27	28	29		23	24	25	26	27	28	29
30	31	1	2	3	4	5	<Sept	30	1	2	3	4	5	6
6	7	8	9	10	11	12	Elul>	7	8	9	10	11	12	13
13	14	15	16	17	18	19		14	15	16	17	18	19	20
20	21	22	23	24	25	26		21	22	23	24	25	26	27
27	28	29	30	1	2	3	Tishri>	28	29	1	2	3	4	5
4	5	6	7	8	9	10	<Oct	6	7	8	9	10	11	12
11	12	13	14	15	16	17		13	14	15	16	17	18	19
18	19	20	21	22	23	24		20	21	22	23	24	25	26
25	26	27	28	29	30	31	Heshvan>	27	28	29	30	1	2	3
1	2	3	4	5	6	7	<Nov	4	5	6	7	8	9	10
8	9	10	11	12	13	14		11	12	13	14	15	16	17
15	16	17	18	19	20	21		18	19	20	21	22	23	24
22	23	24	25	26	27	28	*Kislev>	25	26	27	28	29	1	2
29	30	1	2	3	4	5	<Dec	3	4	5	6	7	8	9
6	7	8	9	10	11	12		10	11	12	13	14	15	16
13	14	15	16	17	18	19		17	18	19	20	21	22	23
20	21	22	23	24	25	26		24	25	26	27	28	1	2
27	28	29	30	31			Tebeth>	3	4	5	6	7		

AD 521/Dionysius' Cycle 0, Year 9/Metonic Year 6/Easter 11 April, Luna 19

S	*M*	*T*	*W*	*T*	*F*	*S*		*S*	*M*	*T*	*W*	*T*	*F*	*S*
					1	2	Tebeth>						8	9
3	4	5	6	7	8	9	<Jan	10	11	12	13	14	15	16
10	11	12	13	14	15	16		17	18	19	20	21	22	23
17	18	19	20	21	22	23		24	25	26	27	28	29	1
24	25	26	27	28	29	30	Shevat>	2	3	4	5	6	7	8
31	1	2	3	4	5	6	<Feb	9	10	11	12	13	14	15
7	8	9	10	11	12	13		16	17	18	19	20	21	22
14	15	16	17	18	19	20		23	24	25	26	27	28	29
21	22	23	24	25	26	27	Adar>	30	1	2	3	4	5	6
28	1	2	3	4	5	6	<Mar	7	8	9	10	11	12	13
7	8	9	10	11	12	13		14	15	16	17	18	19	20
14	15	16	17	18	19	20		21	22	23	24	25	26	27
21	22	23	24	25	26	27	Nisan>	28	29	30	1	2	3	4
28	29	30	31	1	2	3	<April	5	6	7	8	9	10	11
4	5	6	7	8	9	10		12	13	14	15	16	17	18
11	12	13	14	15	16	17		19	20	21	22	23	24	25
18	19	20	21	22	23	24		26	27	28	29	30	1	2
25	26	27	28	29	30	1	Iyar>	3	4	5	6	7	8	9
2	3	4	5	6	7	8	<May	10	11	12	13	14	15	16
9	10	11	12	13	14	15		17	18	19	20	21	22	23
16	17	18	19	20	21	22		24	25	26	27	28	29	1
23	24	25	26	27	28	29	Sivan>	2	3	4	5	6	7	8
30	31	1	2	3	4	5	<June	9	10	11	12	13	14	15
6	7	8	9	10	11	12		16	17	18	19	20	21	22
13	14	15	16	17	18	19		23	24	25	26	27	28	29
20	21	22	23	24	25	26		30	1	2	3	4	5	6
27	28	29	30	1	2	3	Tammuz>	7	8	9	10	11	12	13
4	5	6	7	8	9	10	<July	14	15	16	17	18	19	20
11	12	13	14	15	16	17		21	22	23	24	25	26	27
18	19	20	21	22	23	24		28	29	1	2	3	4	5
25	26	27	28	29	30	31	Ab>	6	7	8	9	10	11	12
1	2	3	4	5	6	7	<August	13	14	15	16	17	18	19
8	9	10	11	12	13	14		20	21	22	23	24	25	26
15	16	17	18	19	20	21		27	28	29	30	1	2	3
22	23	24	25	26	27	28	Elul>	4	5	6	7	8	9	10
29	30	31	1	2	3	4	<Sept	11	12	13	14	15	16	17
5	6	7	8	9	10	11		18	19	20	21	22	23	24
12	13	14	15	16	17	18		25	26	27	28	29	1	2
19	20	21	22	23	24	25	Tishri>	3	4	5	6	7	8	9
26	27	28	29	30	1	2		10	11	12	13	14	15	16
3	4	5	6	7	8	9	<Oct	17	18	19	20	21	22	23
10	11	12	13	14	15	16		24	25	26	27	28	29	30
17	18	19	20	21	22	23	Heshvan>	1	2	3	4	5	6	7
24	25	26	27	28	29	30		8	9	10	11	12	13	14
31	1	2	3	4	5	6	<Nov	15	16	17	18	19	20	21
7	8	9	10	11	12	13		22	23	24	25	26	27	28
14	15	16	17	18	19	20		29	1	2	3	4	5	6
21	22	23	24	25	26	27	Kislev>	7	8	9	10	11	12	13
28	29	30	1	2	3	4	<Dec	14	15	16	17	18	19	20
5	6	7	8	9	10	11		21	22	23	24	25	26	27
12	13	14	15	16	17	18		28	29	30	1	2	3	4
19	20	21	22	23	24	25	Tebeth>	5	6	7	8	9	10	11
26	27	28	29	30	31			12	13	14	15	16	17	

AD 522/Dionysius' Cycle 0, Year 10/Metonic Year 7/Easter 03 April, Luna 21

S	*M*	*T*	*W*	*T*	*F*	*S*		*S*	*M*	*T*	*W*	*T*	*F*	*S*
						1	Tebeth>							18
2	3	4	5	6	7	8	<Jan	19	20	21	22	23	24	25
9	10	11	12	13	14	15		26	27	28	29	1	2	3
16	17	18	19	20	21	22	Shevat>	4	5	6	7	8	9	10
23	24	25	26	27	28	29		11	12	13	14	15	16	17
30	31	1	2	3	4	5	<Feb	18	19	20	21	22	23	24
6	7	8	9	10	11	12		25	26	27	28	29	30	1
13	14	15	16	17	18	19	Adar>	2	3	4	5	6	7	8
20	21	22	23	24	25	26		9	10	11	12	13	14	15
27	28	1	2	3	4	5	<March	16	17	18	19	20	21	22
6	7	8	9	10	11	12		23	24	25	26	27	28	29
13	14	15	16	17	18	19	Nisan>	30	1	2	3	4	5	6
20	21	22	23	24	25	26		7	8	9	10	11	12	13
27	28	29	30	31	1	2		14	15	16	17	18	19	20
3	4	5	6	7	8	9	<April	21	22	23	24	25	26	27
10	11	12	13	14	15	16		28	29	30	1	2	3	4
17	18	19	20	21	22	23	Iyar>	5	6	7	8	9	10	11
24	25	26	27	28	29	30		12	13	14	15	16	17	18
1	2	3	4	5	6	7	<May	19	20	21	22	23	24	25
8	9	10	11	12	13	14		26	27	28	29	1	2	3
15	16	17	18	19	20	21	Sivan>	4	5	6	7	8	9	10
22	23	24	25	26	27	28		11	12	13	14	15	16	17
29	30	31	1	2	3	4	<June	18	19	20	21	22	23	24
5	6	7	8	9	10	11		25	26	27	28	29	30	1
12	13	14	15	16	17	18	Tammuz>	2	3	4	5	6	7	8
19	20	21	22	23	24	25		9	10	11	12	13	14	15
26	27	28	29	30	1	2		16	17	18	19	20	21	22
3	4	5	6	7	8	9	<July	23	24	25	26	27	28	29
10	11	12	13	14	15	16	Ab>	1	2	3	4	5	6	7
17	18	19	20	21	22	23		8	9	10	11	12	13	14
24	25	26	27	28	29	30		15	16	17	18	19	20	21
31	1	2	3	4	5	6	<August	22	23	24	25	26	27	28
7	8	9	10	11	12	13	Elul>	29	30	1	2	3	4	5
14	15	16	17	18	19	20		6	7	8	9	10	11	12
21	22	23	24	25	26	27		13	14	15	16	17	18	19
28	29	30	31	1	2	3		20	21	22	23	24	25	26
4	5	6	7	8	9	10	<Sept	27	28	29	1	2	3	4
11	12	13	14	15	16	17	Tishri>	5	6	7	8	9	10	11
18	19	20	21	22	23	24		12	13	14	15	16	17	18
25	26	27	28	29	30	1		19	20	21	22	23	24	25
2	3	4	5	6	7	8	<Oct	26	27	28	29	30	1	2
9	10	11	12	13	14	15	Heshvan>	3	4	5	6	7	8	9
16	17	18	19	20	21	22		10	11	12	13	14	15	16
23	24	25	26	27	28	29		17	18	19	20	21	22	23
30	31	1	2	3	4	5	<Nov	24	25	26	27	28	29	1
6	7	8	9	10	11	12	Kislev>	2	3	4	5	6	7	8
13	14	15	16	17	18	19		9	10	11	12	13	14	15
20	21	22	23	24	25	26		16	17	18	19	20	21	22
27	28	29	30	1	2	3	<Dec	23	24	25	26	27	28	29
4	5	6	7	8	9	10	Tebeth>	30	1	2	3	4	5	6
11	12	13	14	15	16	17		7	8	9	10	11	12	13
18	19	20	21	22	23	24		14	15	16	17	18	19	20
25	26	27	28	29	30	31		21	22	23	24	25	26	27

AD 523/Dionysius' Cycle 0, Year 11 (emb.)/Metonic Year 8/ Easter 16 April, Luna 15

S	*M*	*T*	*W*	*T*	*F*	*S*	*Tebeth>*	*S*	*M*	*T*	*W*	*T*	*F*	*S*
1	2	3	4	5	6	7	<Jan	28	29	1	2	3	4	5
8	9	10	11	12	13	14	Shevat>	6	7	8	9	10	11	12
15	16	17	18	19	20	21		13	14	15	16	17	18	19
22	23	24	25	26	27	28		20	21	22	23	24	25	26
29	30	31	1	2	3	4	Feb<	27	28	29	30	1	2	3
5	6	7	8	9	10	11	Adar>	4	5	6	7	8	9	10
12	13	14	15	16	17	18		11	12	13	14	15	16	17
19	20	21	22	23	24	25		18	19	10	11	12	13	14
26	27	28	1	2	3	4	<March	25	26	27	28	29	30	1
5	6	7	8	9	10	11	Adar 2>	2	3	4	5	6	7	8
12	13	14	15	16	17	18		9	10	11	12	13	14	15
19	20	21	22	23	24	25		16	17	18	19	20	21	22
26	27	28	29	30	31	1		23	24	25	26	27	28	29
2	3	4	5	6	7	8	<April	1	2	3	4	5	6	7
9	10	11	12	13	14	15	Nisan>	8	9	10	11	12	13	14
16	17	18	19	20	21	22		15	16	17	18	19	20	21
23	24	25	26	27	28	29		22	23	24	25	26	27	28
30	1	2	3	4	5	6	<May	29	30	1	2	3	4	5
7	8	9	10	11	12	13	Iyar>	6	7	8	9	10	11	12
14	15	16	17	18	19	20		13	14	15	16	17	18	19
21	22	23	24	25	26	27		20	21	22	23	24	25	26
28	29	30	31	1	2	3	Sivan>	27	28	29	1	2	3	4
4	5	6	7	8	9	10	<June	5	6	7	8	9	10	11
11	12	13	14	15	16	17		12	13	14	15	16	17	18
18	19	20	21	22	23	24		19	20	21	22	23	24	25
25	26	27	28	29	30	1	Tammuz>	26	27	28	29	30	1	2
2	3	4	5	6	7	8	<July	3	4	5	6	7	8	9
9	10	11	12	13	14	15		10	11	12	13	14	15	16
16	17	18	19	20	21	22		17	18	19	20	21	22	23
23	24	25	26	27	28	29		24	25	26	27	28	29	1
30	31	1	2	3	4	5	<August	2	3	4	5	6	7	8
6	7	8	9	10	11	12	Ab>	9	10	11	12	13	14	15
13	14	15	16	17	18	19		16	17	18	19	20	21	22
20	21	22	23	24	25	26		23	24	25	26	27	28	29
27	28	29	30	31	1	2	Elul>	30	1	2	3	4	5	6
3	4	5	6	7	8	9	<Sept	7	8	9	10	11	12	13
10	11	12	13	14	15	16		14	15	16	17	18	19	20
17	18	19	20	21	22	23		21	22	23	24	25	26	27
24	25	26	27	28	29	30	Tishri>	28	29	1	2	3	4	5
1	2	3	4	5	6	7	<Oct	6	7	8	9	10	11	12
8	9	10	11	12	13	14		13	14	15	16	17	18	19
15	16	17	18	19	20	21		20	21	22	23	24	25	26
22	23	24	25	26	27	28	Heshvan>	27	28	29	30	1	2	3
29	30	31	1	2	3	4	<Nov	4	5	6	7	8	9	10
5	6	7	8	9	10	11		11	12	13	14	15	16	17
12	13	14	15	16	17	18		18	19	20	21	22	23	24
19	20	21	22	23	24	25		25	26	27	28	29	1	2
26	27	28	29	30	1	2	Kislev>	3	4	5	6	7	8	9
3	4	5	6	7	8	9	<Dec	10	11	12	13	14	15	16
10	11	12	13	14	15	16		17	18	19	20	21	22	23
17	18	19	20	21	22	23		24	25	26	27	28	29	30
24	25	26	27	28	29	30	Tebeth>	1	2	3	4	5	6	7
31								8						

AD 524/Dionysius' Cycle 0, Year 12/Metonic Year 9/Easter 07 April, Luna 17

S	*M*	*T*	*W*	*T*	*F*	*S*		*S*	*M*	*T*	*W*	*T*	*F*	*S*
	1	2	3	4	5	6	Tebeth>		9	10	11	12	13	14
7	8	9	10	11	12	13	<Jan	15	16	17	18	19	20	21
14	15	16	17	18	19	20		22	23	24	25	26	27	28
21	22	23	24	25	26	27	Shebat>	29	1	2	3	4	5	6
28	29	30	31	1	2	3		7	8	9	10	11	12	13
4	5	6	7	8	9	10	<Feb	14	15	16	17	18	19	20
11	12	13	14	15	16	17		21	22	23	24	25	26	27
18	19	20	21	22	23	24	Adar>	28	29	30	1	2	3	4
25	26	27	28	29	1	2		5	6	7	8	9	10	11
3	4	5	6	7	8	9	<March	12	13	14	15	16	17	18
10	11	12	13	14	15	16		19	20	21	22	23	24	25
17	18	19	20	21	22	23		26	27	28	29	30	1	2
24	25	26	27	28	29	30	Nisan>	3	4	5	6	7	8	9
31	1	2	3	4	5	6	<April	10	11	12	13	14	15	16
7	8	9	10	11	12	13		17	18	19	20	21	22	23
14	15	16	17	18	19	20		24	25	26	27	28	29	30
21	22	23	24	25	26	27	Iyar>	1	2	3	4	5	6	7
28	29	30	1	2	3	4	<May	8	9	10	11	12	13	14
5	6	7	8	9	10	11		15	16	17	18	19	20	21
12	13	14	15	16	17	18		22	23	24	25	26	27	28
19	20	21	22	23	24	25	Sivan>	29	1	2	3	4	5	6
26	27	28	29	30	31	1		7	8	9	10	11	12	13
2	3	4	5	6	7	8	<June	14	15	16	17	18	19	20
9	10	11	12	13	14	15		21	22	23	24	25	26	27
16	17	18	19	20	21	22	Tammuz>	28	29	30	1	2	3	4
23	24	25	26	27	28	29		5	6	7	8	9	10	11
30	1	2	3	4	5	6	<July	12	13	14	15	16	17	18
7	8	9	10	11	12	13		19	20	21	22	23	24	25
14	15	16	17	18	19	20		26	27	28	29	1	2	3
21	22	23	24	25	26	27	Ab>	4	5	6	7	8	9	10
28	29	30	31	1	2	3		11	12	13	14	15	16	17
4	5	6	7	8	9	10	<August	18	19	20	21	22	23	24
11	12	13	14	15	16	17		25	26	27	28	29	30	1
18	19	20	21	22	23	24	Elul>	2	3	4	5	6	7	8
25	26	27	28	29	30	31		9	10	11	12	13	14	15
1	2	3	4	5	6	7	<Sept	16	17	18	19	20	21	22
8	9	10	11	12	13	14		23	24	25	26	27	28	29
15	16	17	18	19	20	21	Tishri>	1	2	3	4	5	6	7
22	23	24	25	26	27	28		8	9	10	11	12	13	14
29	30	1	2	3	4	5	<Oct	15	16	17	18	19	20	21
6	7	8	9	10	11	12		22	23	24	25	26	27	28
13	14	15	16	17	18	19	Heshvan>	29	30	1	2	3	4	5
20	21	22	23	24	25	26		6	7	8	9	10	11	12
27	28	29	30	31	1	2		13	14	15	16	17	18	19
3	4	5	6	7	8	9	<Nov	20	21	22	23	24	25	26
10	11	12	13	14	15	16	Kislev>	27	28	29	1	2	3	4
17	18	19	20	21	22	23		5	6	7	8	9	10	11
24	25	26	27	28	29	30		12	13	14	15	16	17	18
1	2	3	4	5	6	7	<Dec	19	20	21	22	23	24	25
8	9	10	11	12	13	14		26	27	28	29	1	2	3
15	16	17	18	19	20	21	Tebeth>	4	5	6	7	8	9	10
22	23	24	25	26	27	28		11	12	13	14	15	16	17
29	30	31						18	19	20				

AD 525/Dionysius' Cycle 0, Year 13/Metonic Year 10/Easter 30 March, Luna 20

S	*M*	*T*	*W*	*T*	*F*	*S*		*S*	*M*	*T*	*W*	*T*	*F*	*S*
			1	2	3	4	Tebeth>				21	22	23	24
5	6	7	8	9	10	11	<Jan	25	26	27	28	29	1	2
12	13	14	15	16	17	18	Shevat>	3	4	5	6	7	8	9
19	20	21	22	23	24	25		10	11	12	13	14	15	16
26	27	28	29	30	31	1		17	18	19	20	21	22	23
2	3	4	5	6	7	8	<Feb	24	25	26	27	28	29	30
9	10	11	12	13	14	15	Adar>	1	2	3	4	5	6	7
16	17	18	19	20	21	22		8	9	10	11	12	13	14
23	24	25	26	27	28	1		15	16	17	18	19	20	21
2	3	4	5	6	7	8	<March	22	23	24	25	26	27	28
9	10	11	12	13	14	15	Nisan>	29	30	1	2	3	4	5
16	17	18	19	20	21	22		6	7	8	9	10	11	12
23	24	25	26	27	28	29		13	14	15	16	17	18	19
30	31	1	2	3	4	5	<April	20	21	22	23	24	25	26
6	7	8	9	10	11	12		27	28	29	30	1	2	3
13	14	15	16	17	18	19	Iyar>	4	5	6	7	8	9	10
20	21	22	23	24	25	26		11	12	13	14	15	16	17
27	28	29	30	1	2	3		18	19	20	21	22	23	24
4	5	6	7	8	9	10	<May	25	26	27	28	29	1	2
11	12	13	14	15	16	17	Sivan>	3	4	5	6	7	8	9
18	19	20	21	22	23	24		10	11	12	13	14	15	16
25	26	27	28	29	30	31		17	18	19	20	21	22	23
1	2	3	4	5	6	7	<June	24	25	26	27	28	29	30
8	9	10	11	12	13	14	Tammuz>	1	2	3	4	5	6	7
15	16	17	18	19	20	21		8	9	10	11	12	13	14
22	23	24	25	26	27	28		15	16	17	18	19	20	21
29	30	1	2	3	4	5	<July	22	23	24	25	26	27	28
6	7	8	9	10	11	12	Ab>	29	1	2	3	4	5	6
13	14	15	16	17	18	19		7	8	9	10	11	12	13
20	21	22	23	24	25	26		14	15	16	17	18	19	20
27	28	29	30	31	1	2		21	22	23	24	25	26	27
3	4	5	6	7	8	9	<August	28	29	30	1	2	3	4
10	11	12	13	14	15	16	Elul>	5	6	7	8	9	10	11
17	18	19	20	21	22	23		12	13	14	15	16	17	18
24	25	26	27	28	29	30		19	20	21	22	23	24	25
31	1	2	3	4	5	6	<Sept	26	27	28	29	1	2	3
7	8	9	10	11	12	13	Tishri>	4	5	6	7	8	9	10
14	15	16	17	18	19	20		11	12	13	14	15	16	17
21	22	23	24	25	26	27		18	19	20	21	22	23	24
28	29	30	1	2	3	4	<Oct	25	26	27	28	29	30	1
5	6	7	8	9	10	11	Heshvan>	2	3	4	5	6	7	8
12	13	14	15	16	17	18		9	10	11	12	13	14	15
19	20	21	22	23	24	25		16	17	18	19	20	21	22
26	27	28	29	30	31	1		23	24	25	26	27	28	29
2	3	4	5	6	7	8	<Nov	1	2	3	4	5	6	7
9	10	11	12	13	14	15	Kislev>	8	9	10	11	12	13	14
16	17	18	19	20	21	22		15	16	17	18	19	20	21
23	24	25	26	27	28	29		22	23	24	25	26	27	28
30	1	2	3	4	5	6	<Dec	29	30	1	2	3	4	5
7	8	9	10	11	12	13	Tebeth>	6	7	8	9	10	11	12
14	15	16	17	18	19	20		13	14	15	16	17	18	19
21	22	23	24	25	26	27		20	21	22	23	24	25	26
28	29	30	31				Shevat>	27	28	29	1			

AD 526/Dionysius' Cycle 0, Year 14 (emb.)/Metonic Year 11/Easter 19 April, Luna 21

S	*M*	*T*	*W*	*T*	*F*	*S*		*S*	*M*	*T*	*W*	*T*	*F*	*S*
				1	2	3	Shevat>	S				2	3	4
4	5	6	7	8	9	10	<Jan	5	6	7	8	9	10	11
11	12	13	14	15	16	17		12	13	14	15	16	17	18
18	19	20	21	22	23	24		19	20	21	22	23	24	25
25	26	27	28	29	30	31	Adar>	26	27	28	29	30	1	2
1	2	3	4	5	6	7	<Feb	3	4	5	6	7	8	9
8	9	10	11	12	13	14		10	11	12	13	14	15	16
15	16	17	18	19	20	21		17	18	19	20	21	22	23
22	23	24	25	26	27	28		24	25	26	27	28	29	30
1	2	3	4	5	6	7	<March	1	2	3	4	5	6	7
8	9	10	11	12	13	14	Adar 2>	8	9	10	11	12	13	14
15	16	17	18	19	20	21		15	16	17	18	19	20	21
22	23	24	25	26	27	28		22	23	24	25	26	27	28
29	30	31	1	2	3	4	<April	29	1	2	3	4	5	6
5	6	7	8	9	10	11	Nisan>	7	8	9	10	11	12	13
12	13	14	15	16	17	18		14	15	16	17	18	19	20
19	20	21	22	23	24	25		21	22	23	24	25	26	27
26	27	28	29	30	1	2	Iyar>	28	29	30	1	2	3	4
3	4	5	6	7	8	9	<May	5	6	7	8	9	10	11
10	11	12	13	14	15	16		12	13	14	15	16	17	18
17	18	19	20	21	22	23		19	20	21	22	23	24	25
24	25	26	27	28	29	30	Sivan>	26	27	28	29	1	2	3
31	1	2	3	4	5	6	<June	4	5	6	7	8	9	10
7	8	9	10	11	12	13		11	12	13	14	15	16	17
14	15	16	17	18	19	20		18	19	20	21	22	23	24
21	22	23	24	25	26	27		25	26	27	28	29	30	1
28	29	30	1	2	3	4	<July	2	3	4	5	6	7	8
5	6	7	8	9	10	11	Tammuz>	9	10	11	12	13	14	15
12	13	14	15	16	17	18		16	17	18	19	20	21	22
19	20	21	22	23	24	25		23	24	25	26	27	28	29
26	27	28	29	30	31	1	Ab>	1	2	3	4	5	6	7
2	3	4	5	6	7	8	<August	8	9	10	11	12	13	14
9	10	11	12	13	14	15		15	16	17	18	19	20	21
16	17	18	19	20	21	22		22	23	24	25	26	27	28
23	24	25	26	27	28	29	Elul>	29	30	1	2	3	4	5
30	31	1	2	3	4	5	<Sept	6	7	8	9	10	11	12
6	7	8	9	10	11	12		13	14	15	16	17	18	19
13	14	15	16	17	18	19		20	21	22	23	24	25	26
20	21	22	23	24	25	26	Tishri>	27	28	29	1	2	3	4
27	28	29	30	1	2	3		5	6	7	8	9	10	11
4	5	6	7	8	9	10	<Oct	12	13	14	15	16	17	18
11	12	13	14	15	16	17		19	20	21	22	23	24	25
18	19	20	21	22	23	24		26	27	28	29	30	1	2
25	26	27	28	29	30	31	Heshvan>	3	4	5	6	7	8	9
1	2	3	4	5	6	7	<Nov	10	11	12	13	14	15	16
8	9	10	11	12	13	14		17	18	19	20	21	22	23
15	16	17	18	19	20	21		24	25	26	27	28	29	1
22	23	24	25	26	27	28	Kislev>	2	3	4	5	6	7	8
29	30	1	2	3	4	5	<Dec	9	10	11	12	13	14	15
6	7	8	9	10	11	12		16	17	18	19	20	21	22
13	14	15	16	17	18	19		23	24	25	26	27	28	29
20	21	22	23	24	25	26		1	2	3	4	5	6	7
27	28	29	30	31			Tebeth>	8	9	10	11	12		

AD 527/ Dionysius' Cycle 0, Year 15/Metonic Year 12/Easter 04 April, Luna 17

S	*M*	*T*	*W*	*T*	*F*	*S*		*S*	*M*	*T*	*W*	*T*	*F*	*S*
					1	2	Tebeth>						13	14
3	4	5	6	7	8	9	<Jan	15	16	17	18	19	20	21
10	11	12	13	14	15	16		22	23	24	25	26	27	28
17	18	19	20	21	22	23	Shevat>	29	1	2	3	4	5	6
24	25	26	27	28	29	30		7	8	9	10	11	12	13
31	1	2	3	4	5	6	<Feb	14	15	16	17	18	19	20
7	8	9	10	11	12	13		21	22	23	24	25	26	27
14	15	16	17	18	19	20	Adar>	28	29	30	1	2	3	4
21	22	23	24	25	26	27		5	6	7	8	9	10	11
28	1	2	3	4	5	6	<March	12	13	14	15	16	17	18
7	8	9	10	11	12	13		19	20	21	22	23	24	25
14	15	16	17	18	19	20		26	27	28	29	30	1	2
21	22	23	24	25	26	27	Nisan>	3	4	5	6	7	8	9
28	29	30	31	1	2	3	<April	10	11	12	13	14	15	16
4	5	6	7	8	9	10		17	18	19	20	21	22	23
11	12	13	14	15	16	17		24	25	26	27	28	29	30
18	19	20	21	22	23	24	Iyar>	1	2	3	4	5	6	7
25	26	27	28	29	30	1		8	9	10	11	12	13	14
2	3	4	5	6	7	8	<May	15	16	17	18	19	20	21
9	10	11	12	13	14	15		22	23	24	25	26	27	28
16	17	18	19	20	21	22	Sivan>	29	1	2	3	4	5	6
23	24	25	26	27	28	29		7	8	9	10	11	12	13
30	31	1	2	3	4	5	<June	14	15	16	17	18	19	20
6	7	8	9	10	11	12		21	22	23	24	25	26	27
13	14	15	16	17	18	19	Tammuz>	28	29	30	1	2	3	4
20	21	22	23	24	25	26		5	6	7	8	9	10	11
27	28	29	30	1	2	3	<July	12	13	14	15	16	17	18
4	5	6	7	8	9	10		19	20	21	22	23	24	25
11	12	13	14	15	16	17		26	27	28	29	1	2	3
18	19	20	21	22	23	24	Ab>	4	5	6	7	8	9	10
25	26	27	28	29	30	31		11	12	13	14	15	16	17
1	2	3	4	5	6	7	<August	18	19	20	21	22	23	24
8	9	10	11	12	13	14		25	26	27	28	29	30	1
15	16	17	18	19	20	21	Elul>	2	3	4	5	6	7	8
22	23	24	25	26	27	28		9	10	11	12	13	14	15
29	30	31	1	2	3	4	<Sept	16	17	18	19	20	21	22
5	6	7	8	9	10	11		23	24	25	26	27	28	29
12	13	14	15	16	17	18	Tishri>	1	2	3	4	5	6	7
19	20	21	22	23	24	25		8	9	10	11	12	13	14
26	27	28	29	30	1	2		15	16	17	18	19	20	21
3	4	5	6	7	8	9	<Oct	22	23	24	25	26	27	28
10	11	12	13	14	15	16	Heshvan>	29	30	1	2	3	4	5
17	18	19	20	21	22	23		6	7	8	9	10	11	12
24	25	26	27	28	29	30		13	14	15	16	17	18	19
31	1	2	3	4	5	6	<Nov	20	21	22	23	24	25	26
7	8	9	10	11	12	13	Kislev>	27	28	29	1	2	3	4
14	15	16	17	18	19	20		5	6	7	8	9	10	11
21	22	23	24	25	26	27		12	13	14	15	16	17	18
28	29	30	1	2	3	4		19	20	21	22	23	24	25
5	6	7	8	9	10	11	<Dec	26	27	28	29	30	1	2
12	13	14	15	16	17	18	Tebeth>	3	4	5	6	7	8	9
19	20	21	22	23	24	25		10	11	12	13	14	15	16
26	27	28	29	30	31			17	18	19	20	21	22	

AD 528/Dionysius' Cycle 0, Year 16/Metonic Year 13/Easter 26 March, Luna 19

S	*M*	*T*	*W*	*T*	*F*	*S*		*S*	*M*	*T*	*W*	*T*	*F*	*S*
2	3	4	5	6	7	1/8	Tebeth>	24	25	26	27	28	29	23/1
9	10	11	12	13	14	15	<Jan	2	3	4	5	6	7	8
16	17	18	19	20	21	22	Shevat>	9	10	11	12	13	14	15
23	24	25	26	27	28	29		16	17	18	19	20	21	22
30	31	1	2	3	4	5	<Feb	23	24	25	26	27	28	29
6	7	8	9	10	11	12	Adar>	30	1	2	3	4	5	6
13	14	15	16	17	18	19		7	8	9	10	11	12	13
20	21	22	23	24	25	26		14	15	16	17	18	19	20
27	28	29	1	2	3	4	<March	21	22	23	24	25	26	27
5	6	7	8	9	10	11		28	29	30	1	2	3	4
12	13	14	15	16	17	18	Nisan>	5	6	7	8	9	10	11
19	20	21	22	23	24	25		12	13	14	15	16	17	18
26	27	28	29	30	31	1		19	20	21	22	23	24	25
2	3	4	5	6	7	8	<April	26	27	28	29	30	1	2
9	10	11	12	13	14	15	Iyar>	3	4	5	6	7	8	9
16	17	18	19	20	21	22		10	11	12	13	14	15	16
23	24	25	26	27	28	29		17	18	19	20	21	22	23
30	1	2	3	4	5	6	<May	24	25	26	27	28	29	1
7	8	9	10	11	12	13	Sivan>	2	3	4	5	6	7	8
14	15	16	17	18	19	20		9	10	11	12	13	14	15
21	22	23	24	25	26	27		16	17	18	19	20	21	22
28	29	30	31	1	2	3		23	24	25	26	27	28	29
4	5	6	7	8	9	10	<June	30	1	2	3	4	5	6
11	12	13	14	15	16	17	Tammuz>	7	8	9	10	11	12	13
18	19	20	21	22	23	24		14	15	16	17	18	19	20
25	26	27	28	29	30	1		21	22	23	24	25	26	27
2	3	4	5	6	7	8	<July	28	29	1	2	3	4	5
9	10	11	12	13	14	15	Ab>	6	7	8	9	10	11	12
16	17	18	19	20	21	22		13	14	15	16	17	18	19
23	24	25	26	27	28	29		20	21	22	23	24	25	26
30	31	1	2	3	4	5	<August	27	28	29	30	1	2	3
6	7	8	9	10	11	12	Elul>	4	5	6	7	8	9	10
13	14	15	16	17	18	19		11	12	13	14	15	16	17
20	21	22	23	24	25	26		18	19	20	21	22	23	24
27	28	29	30	31	1	2	Tishri>	25	26	27	28	29	1	2
3	4	5	6	7	8	9	<Sept	3	4	5	6	7	8	9
10	11	12	13	14	15	16		10	11	12	13	14	15	16
17	18	19	20	21	22	23		17	18	19	20	21	22	23
24	25	26	27	28	29	30		24	25	26	27	28	29	30
1	2	3	4	5	6	7	<Oct	1	2	3	4	5	6	7
8	9	10	11	12	13	14	Heshvan>	8	9	10	11	12	13	14
15	16	17	18	19	20	21		15	16	17	18	19	20	21
22	23	24	25	26	27	28		22	23	24	25	26	27	28
29	30	31	1	2	3	4	<Nov	29	1	2	3	4	5	6
5	6	7	8	9	10	11	Kislev>	7	8	9	10	11	12	13
12	13	14	15	16	17	18		14	15	16	17	18	19	20
19	20	21	22	23	24	25		21	22	23	24	25	26	27
26	27	28	29	30	1	2	Tebeth>	28	29	30	1	2	3	4
3	4	5	6	7	8	9	<Dec	5	6	7	8	9	10	11
10	11	12	13	14	15	16		12	13	14	15	16	17	18
17	18	19	20	21	22	23		19	20	21	22	23	24	25
24	25	26	27	28	29	30		26	27	28	29	1	2	3
31							Shevat>	4						

AD 529/Dionysius' Cycle 0, Year 17 (emb.)/Metonic Year 14/Easter 15 April, Luna 20

S	*M*	*T*	*W*	*T*	*F*	*S*		*S*	*M*	*T*	*W*	*T*	*F*	*S*
	1	2	3	4	5	6	Shevat>		5	6	7	8	9	10
7	8	9	10	11	12	13	<Jan	11	12	13	14	15	16	17
14	15	16	17	18	19	20		18	19	20	21	22	23	24
21	22	23	24	25	26	27		25	26	27	28	29	30	1
28	29	30	31	1	2	3	Adar>	2	3	4	5	6	7	8
4	5	6	7	8	9	10	<Feb	9	10	11	12	13	14	15
11	12	13	14	15	16	17		16	17	18	19	20	21	22
18	19	20	21	22	23	24		23	24	25	26	27	28	29
25	26	27	28	1	2	3	Adar 2>	30	1	2	3	4	5	6
4	5	6	7	8	9	10	<March	7	8	9	10	11	12	13
11	12	13	14	15	16	17		14	15	16	17	18	19	20
18	19	20	21	22	23	24		21	22	23	24	25	26	27
25	26	27	28	29	30	31	Nisan>	28	29	1	2	3	4	5
1	2	3	4	5	6	7	<April	6	7	8	9	10	11	12
8	9	10	11	12	13	14		13	14	15	16	17	18	19
15	16	17	18	19	20	21		20	21	22	23	24	25	26
22	23	24	25	26	27	28	Iyar>	27	28	29	30	1	2	3
29	30	1	2	3	4	5	<May	4	5	6	7	8	9	10
6	7	8	9	10	11	12		11	12	13	14	15	16	17
13	14	15	16	17	18	19		18	19	20	21	22	23	24
20	21	22	23	24	25	26		25	26	27	28	29	1	2
27	28	29	30	31	1	2	Sivan>	3	4	5	6	7	8	9
3	4	5	6	7	8	9	<June	10	11	12	13	14	15	16
10	11	12	13	14	15	16		17	18	19	20	21	22	23
17	18	19	20	21	22	23		24	25	26	27	28	29	30
24	25	26	27	28	29	30	Tammuz>	1	2	3	4	5	6	7
1	2	3	4	5	6	7	<July	8	9	10	11	12	13	14
8	9	10	11	12	13	14		15	16	17	18	19	20	21
15	16	17	18	19	20	21		22	23	24	25	26	27	28
22	23	24	25	26	27	28	Ab>	29	1	2	3	4	5	6
29	30	31	1	2	3	4	<August	7	8	9	10	11	12	13
5	6	7	8	9	10	11		14	15	16	17	18	19	20
12	13	14	15	16	17	18		21	22	23	24	25	26	27
19	20	21	22	23	24	25	Elul>	28	29	30	1	2	3	4
26	27	28	29	30	31	1		5	6	7	8	9	10	11
2	3	4	5	6	7	8	<Sept	12	13	14	15	16	17	18
9	10	11	12	13	14	15		19	20	21	22	23	24	25
16	17	18	19	20	21	22		26	27	28	29	1	2	3
23	24	25	26	27	28	29	Tishri>	4	5	6	7	8	9	10
30	1	2	3	4	5	6	<Oct	11	12	13	14	15	16	17
7	8	9	10	11	12	13		18	19	20	21	22	23	24
14	15	16	17	18	19	20		25	26	27	28	29	30	1
21	22	23	24	25	26	27	Heshvan>	2	3	4	5	6	7	8
28	29	30	31	1	2	3		9	10	11	12	13	14	15
4	5	6	7	8	9	10	<Nov	16	17	18	19	20	21	22
11	12	13	14	15	16	17		23	24	25	26	27	28	29
18	19	20	21	22	23	24	Kislev>	1	2	3	4	5	6	7
25	26	27	28	29	30	1		8	9	10	11	12	13	14
2	3	4	5	6	7	8	<Dec	15	16	17	18	19	20	21
9	10	11	12	13	14	15		22	23	24	25	26	27	28
16	17	18	19	20	21	22		29	1	2	3	4	5	6
23	24	25	26	27	28	29	Tebeth>	7	8	9	10	11	12	13
30	31							14	15					

AD 530/Dionysius' Cycle 0, Year 18/Metonic Year 15/Easter 31 March, Luna 16

S	*M*	*T*	*W*	*T*	*F*	*S*		*S*	*M*	*T*	*W*	*T*	*F*	*S*
		1	2	3	4	5	Tebeth>			16	17	18	19	20
6	7	8	9	10	11	12	<Jan	21	22	23	24	25	26	27
13	14	15	16	17	18	19		28	29	1	2	3	4	5
20	21	22	23	24	25	26	Shebat>	6	7	8	9	10	11	12
27	28	29	30	31	1	2		13	14	15	16	17	18	19
3	4	5	6	7	8	9	<Feb	20	21	22	23	24	25	26
10	11	12	13	14	15	16		27	28	29	30	1	2	3
17	18	19	20	21	22	23	Adar>	4	5	6	7	8	9	10
24	25	26	27	28	1	2		11	12	13	14	15	16	17
3	4	5	6	7	8	9	<March	18	19	20	21	22	23	24
10	11	12	13	14	15	16		25	26	27	28	29	30	1
17	18	19	20	21	22	23	Nisan>	2	3	4	5	6	7	8
24	25	26	27	28	29	30		9	10	11	12	13	14	15
31	1	2	3	4	5	6	<April	16	17	18	19	20	21	22
7	8	9	10	11	12	13		23	24	25	26	27	28	29
14	15	16	17	18	19	20	Iyar>	30	1	2	3	4	5	6
21	22	23	24	25	26	27		7	8	9	10	11	12	13
28	29	30	1	2	3	4		14	15	16	17	18	19	20
5	6	7	8	9	10	11	<May	21	22	23	24	25	26	27
12	13	14	15	16	17	18		28	29	1	2	3	4	5
19	20	21	22	23	24	25	Sivan>	6	7	8	9	10	11	12
26	27	28	29	30	31	1		13	14	15	16	17	18	19
2	3	4	5	6	7	8	<June	20	21	22	23	24	25	26
9	10	11	12	13	14	15		27	28	29	30	1	2	3
16	17	18	19	20	21	22	Tammuz>	4	5	6	7	8	9	10
23	24	25	26	27	28	29		11	12	13	14	15	16	17
30	1	2	3	4	5	6	<July	18	19	20	21	22	23	24
7	8	9	10	11	12	13		25	26	27	28	29	1	2
14	15	16	17	18	19	20	Ab>	3	4	5	6	7	8	9
21	22	23	24	25	26	27		10	11	12	13	14	15	16
28	29	30	31	1	2	3		17	18	19	20	21	22	23
4	5	6	7	8	9	10	<August	24	25	26	27	28	29	30
11	12	13	14	15	16	17	Elul>	1	2	3	4	5	6	7
18	19	20	21	22	23	24		8	9	10	11	12	13	14
25	26	27	28	29	30	31		15	16	17	18	19	20	21
1	2	3	4	5	6	7	<Sept	22	23	24	25	26	27	28
8	9	10	11	12	13	14	Tishri>	29	1	2	3	4	5	6
15	16	17	18	29	20	21		7	8	9	10	11	12	13
22	23	24	25	26	27	28		14	15	16	17	18	19	20
29	30	1	2	3	4	5	<Oct	21	22	23	24	25	26	27
6	7	8	9	10	11	12		28	29	30	1	2	3	4
13	14	15	16	17	18	19	Heshvan>	5	6	7	8	9	10	11
20	21	22	23	24	25	26		12	13	14	15	16	17	18
27	28	29	30	31	1	2		19	20	21	22	23	24	25
3	4	5	6	7	8	9	<Nov	26	27	28	29	1	2	3
10	11	12	13	14	15	16	Kislev>	4	5	6	7	8	9	10
17	18	19	20	21	22	23		11	12	13	14	15	16	17
24	25	26	27	28	29	30		18	19	20	21	22	23	24
1	2	3	4	5	6	7	<Dec	25	26	27	28	29	30	1
8	9	10	11	12	13	14	Tebeth>	2	3	4	5	6	7	8
15	16	17	18	19	20	21		9	10	11	12	13	14	15
22	23	24	25	26	27	28		16	17	18	19	20	21	22
29	30	31						23	24	25				

AD 531/Dionysius' Cycle 0, Year 19 (emb.)/Metonic Year 16/Easter 20 April, Luna 17

S	*M*	*T*	*W*	*T*	*F*	*S*		*S*	*M*	*T*	*W*	*T*	*F*	*S*
			1	2	3	4	Tebeth>				26	27	28	29
5	6	7	8	9	10	11	<Jan	1	2	3	4	5	6	7
12	13	14	15	16	17	18	Shevat>	8	9	10	11	12	13	14
19	20	21	22	23	24	25		15	16	17	18	19	20	21
26	27	28	29	30	31	1		22	23	24	25	26	27	28
2	3	4	5	6	7	8	<Feb	29	30	1	2	3	4	5
9	10	11	12	13	14	15	Adar>	6	7	8	9	10	11	12
16	17	18	19	20	21	22		13	14	15	16	17	18	19
23	24	25	26	27	28	1		20	21	22	23	24	25	26
2	3	4	5	6	7	8	<March	27	28	29	30	1	2	3
9	10	11	12	13	14	15	Adar 2>	4	5	6	7	8	9	10
16	17	18	19	20	21	22		11	12	13	14	15	16	17
23	24	25	26	27	28	29		18	19	20	21	22	23	24
30	31	1	2	3	4	5	Nisan>	25	26	27	28	29	1	2
6	7	8	9	10	11	12	<April	3	4	5	6	7	8	9
13	14	15	16	17	18	19		10	11	12	13	14	15	16
20	21	22	23	24	25	26		17	18	19	20	21	22	23
27	28	29	30	1	2	3		24	25	26	27	28	29	30
4	5	6	7	8	9	10	<May	1	2	3	4	5	6	7
11	12	13	14	15	16	17	Iyar>	8	9	10	11	12	13	14
18	19	20	21	22	23	24		15	16	17	18	19	20	21
25	26	27	28	29	30	31		22	23	24	25	26	27	28
1	2	3	4	5	6	7	<June	29	1	2	3	4	5	6
8	9	10	11	12	13	14	Sivan>	7	8	9	10	11	12	13
15	16	17	18	19	20	21		14	15	16	17	18	19	20
22	23	24	25	26	27	28		21	22	23	24	25	26	27
29	30	1	2	3	4	5	<July	28	29	30	1	2	3	4
6	7	8	9	10	11	12	Tammuz>	5	6	7	8	9	10	11
13	14	15	16	17	18	19		12	13	14	15	16	17	18
20	21	22	23	24	25	26		19	20	21	22	23	24	25
27	28	29	30	31	1	2	Ab>	26	27	28	29	1	2	3
3	4	5	6	7	8	9	<August	4	5	6	7	8	9	10
10	11	12	13	14	15	16		11	12	13	14	15	16	17
17	18	19	20	21	22	23		18	19	20	21	22	23	24
24	25	26	27	28	29	30		25	26	27	28	29	30	1
31	1	2	3	4	5	6	<Sept	2	3	4	5	6	7	8
7	8	9	10	11	12	13	Elul>	9	10	11	12	13	14	15
14	15	16	17	18	19	20		16	17	18	19	20	21	22
21	22	23	24	25	26	27		23	24	25	26	27	28	29
28	29	30	1	2	3	4	Tishri>	1	2	3	4	5	6	7
5	6	7	8	9	10	11	<Oct	8	9	10	11	12	13	14
12	13	14	15	16	17	18		15	16	17	18	19	20	21
19	20	21	22	23	24	25		22	23	24	25	26	27	28
26	27	28	29	30	31	1	Heshvan>	29	30	1	2	3	4	5
2	3	4	5	6	7	8	<Nov	6	7	8	9	10	11	12
9	10	11	12	13	14	15		13	14	15	16	17	18	19
16	17	18	19	20	21	22		20	21	22	23	24	25	26
23	24	25	26	27	28	29	Kislev>	27	28	29	1	2	3	4
30	1	2	3	4	5	6	<Dec	5	6	7	8	9	10	11
7	8	9	10	11	12	13		12	13	14	15	16	17	18
14	15	16	17	18	19	20		19	20	21	22	23	24	25
21	22	23	24	25	26	27		26	27	28	29	1	2	3
28	29	30	31				Tebeth>	4	5	6	7			

AD 532/Dionysius' Cycle I, Year 1/Metonic Year 17/Easter 11 April, Luna 20

S	*M*	*T*	*W*	*T*	*F*	*S*		*S*	*M*	*T*	*W*	*T*	*F*	*S*
				1	2	3						8	9	10
4	5	6	7	8	9	10	<Jan	11	12	13	14	15	16	17
11	12	13	14	15	16	17		18	19	20	21	22	23	24
18	19	20	21	22	23	24		25	26	27	28	29	1	2
25	26	27	28	29	30	31	Shevat>	3	4	5	6	7	8	9
1	2	3	4	5	6	7	<Feb	10	11	12	13	14	15	16
8	9	10	11	12	13	14		17	18	19	20	21	22	23
15	16	17	18	19	20	21		24	25	26	27	28	29	30
22	23	24	25	26	27	28	Adar>	1	2	3	4	5	6	7
29	1	2	3	4	5	6	<March	8	9	10	11	12	13	14
7	8	9	10	11	12	13		15	16	17	18	19	20	21
14	15	16	17	18	19	20		22	23	24	25	26	27	28
21	22	23	24	25	26	27	Nisan>	29	30	1	2	3	4	5
28	29	30	31	1	2	3	<April	6	7	8	9	10	11	12
4	5	6	7	8	9	10		13	14	15	16	17	18	19
11	12	13	14	15	16	17		20	21	22	23	24	25	26
18	19	20	21	22	23	24		27	28	29	30	1	2	3
25	26	27	28	29	30	1	Iyar>	4	5	6	7	8	9	10
2	3	4	5	6	7	8	<May	11	12	13	14	15	16	17
9	10	11	12	13	14	15		18	19	20	21	22	23	24
16	17	18	19	20	21	22		25	26	27	28	29	1	2
23	24	25	26	27	28	29	Sivan>	3	4	5	6	7	8	9
30	31	1	2	3	4	5	<June	10	11	12	13	14	15	16
6	7	8	9	10	11	12		17	18	19	20	21	22	23
13	14	15	16	17	18	19		24	25	26	27	28	29	30
20	21	22	23	24	25	26	Tammuz>	1	2	3	4	5	6	7
27	28	29	30	1	2	3	<July	8	9	10	11	12	13	14
4	5	6	7	8	9	10		15	16	17	18	19	20	21
11	12	13	14	15	16	17		22	23	24	25	26	27	28
18	19	20	21	22	23	24	Ab>	29	1	2	3	4	5	6
25	26	27	28	29	30	31		7	8	9	10	11	12	13
1	2	3	4	5	6	7	<August	14	15	16	17	18	19	20
8	9	10	11	12	13	14		21	22	23	24	25	26	27
15	16	17	18	19	20	21	Elul>	28	29	30	1	2	3	4
22	23	24	25	26	27	28		5	6	7	8	9	10	11
29	30	31	1	2	3	4	<Sept	12	13	14	15	16	17	18
5	6	7	8	9	10	11		19	20	21	22	23	24	25
12	13	14	15	16	17	18		26	27	28	29	1	2	3
19	20	21	22	23	24	25	Tishri>	4	5	6	7	8	9	10
26	27	28	29	30	1	2		11	12	13	14	15	16	17
3	4	5	6	7	8	9	<Oct	18	19	20	21	22	23	24
10	11	12	13	14	15	16		25	26	27	28	29	30	1
17	18	19	20	21	22	23	Heshvan>	2	3	4	5	6	7	8
24	25	26	27	28	29	30		9	10	11	12	13	14	15
31	1	2	3	4	5	6	<Nov	16	17	18	19	20	21	22
7	8	9	10	11	12	13		23	24	25	26	27	28	29
14	15	16	17	18	19	20	Kislev>	1	2	3	4	5	6	7
21	22	23	24	25	26	27		8	9	10	11	12	13	14
28	29	30	1	2	3	4	<Dec	15	16	17	18	19	20	21
5	6	7	8	9	10	11		22	23	24	25	26	27	28
12	13	14	15	16	17	18	Tebeth>	29	1	2	3	4	5	6
19	20	21	22	23	24	25		7	8	9	10	11	12	13
26	27	28	29	30	31			14	15	16	17	18	19	

AD 533/Dionysius' Cycle I, Year 2/Metonic Year 18/Easter 27 March, Luna 16

S	*M*	*T*	*W*	*T*	*F*	*S*		*S*	*M*	*T*	*W*	*T*	*F*	*S*
						1	Tebeth>							20
2	3	4	5	6	7	8	<Jan	21	22	23	24	25	26	27
9	10	11	12	13	14	15	Shevat>	28	29	1	2	3	4	5
16	17	18	19	20	21	22		6	7	8	9	10	11	12
23	24	25	26	27	28	29		13	14	15	16	17	18	19
30	31	1	2	3	4	5	<Feb	20	21	22	23	24	25	26
6	7	8	9	10	11	12		27	28	29	30	1	2	3
13	14	15	16	17	18	19	Adar>	4	5	6	7	8	9	10
20	21	22	23	24	25	26		11	12	13	14	15	16	17
27	28	1	2	3	4	5	<March	18	19	20	21	22	23	24
6	7	8	9	10	11	12		25	26	27	28	29	30	1
13	14	15	16	17	18	19	Nisan>	2	3	4	5	6	7	8
20	21	22	23	24	25	26		9	10	11	12	13	14	15
27	28	29	30	31	1	2	<April	16	17	18	19	20	21	22
3	4	5	6	7	8	9		23	24	25	26	27	28	29
10	11	12	13	14	15	16	Iyar>	30	1	2	3	4	5	6
17	18	19	20	21	22	23		7	8	9	10	11	12	13
24	25	26	27	28	29	30		14	15	16	17	18	19	20
1	2	3	4	5	6	7	<May	21	22	23	24	25	26	27
8	9	10	11	12	13	14		28	29	1	2	3	4	5
15	16	17	18	19	20	21	Sivan>	6	7	8	9	10	11	12
22	23	24	25	26	27	28		13	14	15	16	17	18	19
29	30	31	1	2	3	4	<June	20	21	22	23	24	25	26
5	6	7	8	9	10	11		27	28	29	30	1	2	3
12	13	14	15	16	17	18	Tammuz>	4	5	6	7	8	9	10
19	20	21	22	23	24	25		11	12	13	14	15	16	17
26	27	28	29	30	1	2		18	19	20	21	22	23	24
3	4	5	6	7	8	9	<July	25	26	27	28	29	1	2
10	11	12	13	14	15	16	Ab>	3	4	5	6	7	8	9
17	18	19	20	21	22	23		10	11	12	13	14	15	16
24	25	26	27	28	29	30		17	18	19	20	21	22	23
31	1	2	3	4	5	6	<August	24	25	26	27	28	29	30
7	8	9	10	11	12	13	Elul>	1	2	3	4	5	6	7
14	15	16	17	18	19	20		8	9	10	11	12	13	14
21	22	23	24	25	26	27		15	16	17	18	19	20	21
28	29	30	31	1	2	3	<Sept	22	23	24	25	26	27	28
4	5	6	7	8	9	10	Tishri>	29	1	2	3	4	5	6
11	12	13	14	15	16	17		7	8	9	10	11	12	13
18	19	20	21	22	23	24		14	15	16	17	18	19	20
25	26	27	28	29	30	1		21	22	23	24	25	26	27
2	3	4	5	6	7	8	<Oct	28	29	30	1	2	3	4
9	10	11	12	13	14	15	Heshvan>	5	6	7	8	9	10	11
16	17	18	19	20	21	22		12	13	14	15	16	17	18
23	24	25	26	27	28	29		19	20	21	22	23	24	25
30	31	1	2	3	4	5	<Nov	26	27	28	29	30	1	2
6	7	8	9	10	11	12	Kislev>	3	4	5	6	7	8	9
13	14	15	16	17	18	19		10	11	12	13	14	15	16
20	21	22	23	24	25	26		17	18	19	20	21	22	23
27	28	29	30	1	2	3	<Dec	25	26	27	28	29	30	1
4	5	6	7	8	9	10	Tebeth>	2	3	4	5	6	7	8
11	12	13	14	15	16	17		9	10	11	12	13	14	15
18	19	20	21	22	23	24		16	17	18	19	20	21	22
25	26	27	28	29	30	31		23	24	25	26	27	28	29

AD 534/Dionysius' Cycle I, Year 3 (emb.)/Metonic Year 19/Easter 16 April, Luna 17

S	*M*	*T*	*W*	*T*	*F*	*S*		*S*	*M*	*T*	*W*	*T*	*F*	*S*
1	2	3	4	5	6	7	Shevat>	1	2	3	4	5	6	7
8	9	10	11	12	13	14	<Jan	8	9	10	11	12	13	14
15	16	17	18	19	20	21		15	16	17	18	19	20	21
22	23	24	25	26	27	28		22	23	24	25	26	27	28
29	30	31	1	2	3	4	<Feb	29	30	1	2	3	4	5
5	6	7	8	9	10	11	Adar>	6	7	8	9	10	11	12
12	13	14	15	16	17	18		13	14	15	16	17	18	19
19	20	21	22	23	24	25		20	21	22	23	24	25	26
26	27	28	1	2	3	4	<March	27	28	29	30	1	2	3
5	6	7	8	9	10	11	Adar 2>	4	5	6	7	8	9	10
12	13	14	15	16	17	18		11	12	13	14	15	16	17
19	20	21	22	23	24	25		18	19	20	21	22	23	24
26	27	28	29	30	31	1		25	26	27	28	29	1	2
2	3	4	5	6	7	8	<April	3	4	5	6	7	8	9
9	10	11	12	13	14	15		10	11	12	13	14	15	16
16	17	18	19	20	21	22		17	18	19	20	21	22	23
23	24	25	26	27	28	29		24	25	26	27	28	29	30
30	1	2	3	4	5	6	<May	1	2	3	4	5	6	7
7	8	9	10	11	12	13	Iyar>	8	9	10	11	12	13	14
14	15	16	17	18	19	20		15	16	17	18	19	20	21
21	22	23	24	25	26	27		22	23	24	25	26	27	28
28	29	30	31	1	2	3	Sivan>	29	1	2	3	4	5	6
4	5	6	7	8	9	10	<June	7	8	9	10	11	12	13
11	12	13	14	15	16	17		14	15	16	17	18	19	20
18	19	20	21	22	23	24		21	22	23	24	25	26	27
25	26	27	28	29	30	1	Tammuz>	28	29	30	1	2	3	4
2	3	4	5	6	7	8	<July	5	6	7	8	9	10	11
9	10	11	12	13	14	15		12	13	14	15	16	17	18
16	17	18	19	20	21	22		19	20	21	22	23	24	25
23	24	25	26	27	28	29		26	27	28	29	1	2	3
30	31	1	2	3	4	5	Ab>	4	5	6	7	8	9	10
6	7	8	9	10	11	12	<August	11	12	13	14	15	16	17
13	14	15	16	17	18	19		18	19	20	21	22	23	24
20	21	22	23	24	25	26		25	26	27	28	29	30	1
27	28	29	30	31	1	2	Elul>	2	3	4	5	6	7	8
3	4	5	6	7	8	9	<Sept	9	10	11	12	13	14	15
10	11	12	13	14	15	16		16	17	18	19	20	21	22
17	18	19	20	21	22	23		23	24	25	26	27	28	29
24	25	26	27	28	29	30	Tishri>	1	2	3	4	5	6	7
1	2	3	4	5	6	7	<Oct	8	9	10	11	12	13	14
8	9	10	11	12	13	14		15	16	17	18	19	20	21
15	16	17	18	19	20	21		22	23	24	25	26	27	28
22	23	24	25	26	27	28		29	30	1	2	3	4	5
29	30	31	1	2	3	4	Heshvan>	6	7	8	9	10	11	12
5	6	7	8	9	10	11	<Nov	13	14	15	16	17	18	19
12	13	14	15	16	17	18		20	21	22	23	24	25	26
19	20	21	22	23	24	25	Kislev>	27	28	29	1	2	3	4
26	27	28	29	30	1	2		5	6	7	8	9	10	11
3	4	5	6	7	8	9	<Dec	12	13	14	15	16	17	18
10	11	12	13	14	15	16		19	20	21	22	23	24	25
17	18	19	20	21	22	23		26	27	28	29	1	2	3
24	25	26	27	28	29	30	Tebeth>	4	5	6	7	8	9	10
31								11						

AD 535/Dionysius' Cycle I, Year 4/Metonic Year 1/Easter 08 April, Luna 20

S	*M*	*T*	*W*	*T*	*F*	*S*		*S*	*M*	*T*	*W*	*T*	*F*	*S*
	1	2	3	4	5	6	Tebeth>		12	13	14	15	16	17
7	8	9	10	11	12	13	<Jan	18	19	20	21	22	23	24
14	15	16	17	18	19	20		25	26	27	28	29	1	2
21	22	23	24	25	26	27	Shevat>	3	4	5	6	7	8	9
28	29	30	31	1	2	3		10	11	12	13	14	15	16
4	5	6	7	8	9	10	<Feb	17	18	19	20	21	22	23
11	12	13	14	15	16	17		24	25	26	27	28	29	30
18	19	20	21	22	23	24	Adar>	1	2	3	4	5	6	7
25	26	27	28	1	2	3		8	9	10	11	12	13	14
4	5	6	7	8	9	10	<March	15	16	17	18	19	20	21
11	12	13	14	15	16	17		22	23	24	25	26	27	28
18	19	20	21	22	23	24	Nisan>	29	30	1	2	3	4	5
25	26	27	28	29	30	31		6	7	8	9	10	11	12
1	2	3	4	5	6	7	<April	13	14	15	16	17	18	19
8	9	10	11	12	13	14		20	21	22	23	24	25	26
15	16	17	18	19	20	21		27	28	29	30	1	2	3
22	23	24	25	26	27	28	Iyar>	4	5	6	7	8	9	10
29	30	1	2	3	4	5	<May	11	12	13	14	15	16	17
6	7	8	9	10	11	12		18	19	20	21	22	23	24
13	14	15	16	17	18	19		25	26	27	28	29	1	2
20	21	22	23	24	25	26	Sivan>	3	4	5	6	7	8	9
27	28	29	30	31	1	2		10	11	12	13	14	15	16
3	4	5	6	7	8	9	<June	17	18	19	20	21	22	23
10	11	12	13	14	15	16		24	25	26	27	28	29	30
17	18	19	20	21	22	23	Tammuz>	1	2	3	4	5	6	7
24	25	26	27	28	29	30		8	9	10	11	12	13	14
1	2	3	4	5	6	7	<July	15	16	17	18	19	20	21
8	9	10	11	12	13	14		22	23	24	25	26	27	28
15	16	17	18	19	20	21	Ab>	29	1	2	3	4	5	6
22	23	24	25	26	27	28		7	8	9	10	11	12	13
29	30	31	1	2	3	4		14	15	16	17	18	19	20
5	6	7	8	9	10	11	<August	21	22	23	24	25	26	27
12	13	14	15	16	17	18		28	29	30	1	2	3	4
19	20	21	22	23	24	25	Elul>	5	6	7	8	9	10	11
26	27	28	29	30	31	1		12	13	14	15	16	17	18
2	3	4	5	6	7	8	<Sept	19	20	21	22	23	24	25
9	10	11	12	13	14	15		26	27	28	29	1	2	3
16	17	18	19	20	21	22	Tishri>	4	5	6	7	8	9	10
23	24	25	26	27	28	29		11	12	13	14	15	16	17
30	1	2	3	4	5	6	<Oct	18	19	20	21	22	23	24
7	8	9	10	11	12	13		25	26	27	28	29	30	1
14	15	16	17	18	19	20	Heshvan>	2	3	4	5	6	7	8
21	22	23	24	25	26	27		9	10	11	12	13	14	15
28	29	30	31	1	2	3		16	17	18	19	20	21	22
4	5	6	7	8	9	10	<Nov	23	24	25	26	27	28	29
11	12	13	14	15	16	17	Kislev>	1	2	3	4	5	6	7
18	19	20	21	22	23	24		8	9	10	11	12	13	14
25	26	27	28	29	30	1		15	16	17	18	19	20	21
2	3	4	5	6	7	8	<Dec	22	23	24	25	26	27	28
9	10	11	12	13	14	15		29	1	2	3	4	5	6
16	17	18	19	20	21	22		7	8	9	10	11	12	13
23	24	25	26	27	28	29		14	15	16	17	18	19	20
30	31							21	22					

AD 536/Dionysius' Cycle I, Year 5/Metonic Year 2/Easter 23 March, Luna 15

S	*M*	*T*	*W*	*T*	*F*	*S*		*S*	*M*	*T*	*W*	*T*	*F*	*S*
		1	2	3	4	5	Tebeth>			22	23	24	25	26
6	7	8	9	10	11	12	<Jan	27	28	29	1	2	3	4
13	14	15	16	17	18	19	Shebat>	5	6	7	8	9	10	11
20	21	22	23	24	25	26		12	13	14	15	16	17	18
27	28	29	30	31	1	2		19	20	21	22	23	24	25
3	4	5	6	7	8	9	<Feb	26	27	28	29	30	1	2
10	11	12	13	14	15	16	Adar>	3	4	5	6	7	8	9
17	18	19	20	21	22	23		10	11	12	13	14	15	16
24	25	26	27	28	29	1		17	18	19	20	21	22	23
2	3	4	5	6	7	8	<March	24	25	26	27	28	29	30
9	10	11	12	13	14	15	Nisan>	1	2	3	4	5	6	7
16	17	18	19	20	21	22		8	9	10	11	12	13	14
23	24	25	26	27	28	29		15	16	17	18	19	20	21
30	31	1	2	3	4	5	<April	22	23	24	25	26	27	28
6	7	8	9	10	11	12		29	30	1	2	3	4	5
13	14	15	16	17	18	19	Iyar>	6	7	8	9	10	11	12
20	21	22	23	24	25	26		13	14	15	16	17	18	19
27	28	29	30	1	2	3		20	21	22	23	24	25	26
4	5	6	7	8	9	10	<May	27	28	29	1	2	3	4
11	12	13	14	15	16	17	Sivan>	5	6	7	8	9	10	11
18	19	20	21	22	23	24		12	13	14	15	16	17	18
25	26	27	28	29	30	31		19	20	21	22	23	24	25
1	2	3	4	5	6	7	<June	26	27	28	29	30	1	2
8	9	10	11	12	13	14	Tammuz>	3	4	5	6	7	8	9
15	16	17	18	19	20	21		10	11	12	13	14	15	16
22	23	24	25	26	27	28		17	18	19	20	21	22	23
29	30	1	2	3	4	5	<July	24	25	26	27	28	29	1
6	7	8	9	10	11	12	Ab>	2	3	4	5	6	7	8
13	14	15	16	17	18	19		9	10	11	12	13	14	15
20	21	22	23	24	25	26		16	17	18	19	20	21	22
27	28	29	30	31	1	2		23	24	25	26	27	28	29
3	4	5	6	7	8	9	<August	30	1	2	3	4	5	6
10	11	12	13	14	15	16	Elul>	7	8	9	10	11	12	13
17	18	19	20	21	22	23		14	15	16	17	18	19	20
24	25	26	27	28	29	30		21	22	23	24	25	26	27
31	1	2	3	4	5	6	<Sept	28	29	1	2	3	4	5
7	8	9	10	11	12	13	Tishri>	6	7	8	9	10	11	12
14	15	16	17	18	19	20		13	14	15	16	17	18	19
21	22	23	24	25	26	27		20	21	22	23	24	25	26
28	29	30	1	2	3	4	<Oct	27	28	29	30	1	2	3
5	6	7	8	9	10	11	Heshvan>	4	5	6	7	8	9	10
12	13	14	15	16	17	18		11	12	13	14	15	16	17
19	20	21	22	23	24	25		18	19	20	21	22	23	24
26	27	28	29	30	31	1	<Nov	25	26	27	28	29	1	2
2	3	4	5	6	7	8	Kislev>	3	4	5	6	7	8	9
9	10	11	12	13	14	15		10	11	12	13	14	15	16
16	17	18	19	20	21	22		17	18	19	20	21	22	23
23	24	25	26	27	28	29		24	25	26	27	28	29	30
30	1	2	3	4	5	6	Tebeth>	1	2	3	4	5	6	7
7	8	9	10	11	12	13	<Dec	8	9	10	11	12	13	14
14	15	16	17	18	19	20		15	16	17	18	19	20	21
21	22	23	24	25	26	27		22	23	24	25	26	27	28
28	29	30	31				Shebat>	29	1	2	3			

AD 537/Dionysius' Cycle I, Year 6 (emb.)/Metonic Year 3/Easter 12 April, Luna 16

S	*M*	*T*	*W*	*T*	*F*	*S*		*S*	*M*	*T*	*W*	*T*	*F*	*S*
				1	2	3	Shebat>					4	5	6
4	5	6	7	8	9	10	<Jan	7	8	9	10	11	12	13
11	12	13	14	15	16	17		14	15	16	17	18	19	20
18	19	20	21	22	23	24		21	22	23	24	25	26	27
25	26	27	28	29	30	31	Adar>	28	29	30	1	2	3	4
1	2	3	4	5	6	7	<Feb	5	6	7	8	9	10	11
8	9	10	11	12	13	14		12	13	14	15	16	17	18
15	16	17	18	19	20	21		19	20	21	22	23	24	25
22	23	24	25	26	27	28	Adar 2>	26	27	28	29	30	1	2
1	2	3	4	5	6	7	<March	3	4	5	6	7	8	9
8	9	10	11	12	13	14		10	11	12	13	14	15	16
15	16	17	18	19	20	21		17	18	19	20	21	22	23
22	23	24	25	26	27	28		24	25	26	27	28	29	1
29	30	31	1	2	3	4	Nisan>	2	3	4	5	6	7	8
5	6	7	8	9	10	11	<April	9	10	11	12	13	14	15
12	13	14	15	16	17	18		16	17	18	19	20	21	22
19	20	21	22	23	24	25		23	24	25	26	27	28	29
26	27	28	29	30	1	2	Iyar>	30	1	2	3	4	5	6
3	4	5	6	7	8	9	<May	7	8	9	10	11	12	13
10	11	12	13	14	15	16		14	15	16	17	18	19	20
17	18	19	20	21	22	23		21	22	23	24	25	26	27
24	25	26	27	28	29	30	Sivan>	28	29	1	2	3	4	5
31	1	2	3	4	5	6	<June	6	7	8	9	10	11	12
7	8	9	10	11	12	13		13	14	15	16	17	18	19
14	15	16	17	18	19	20		20	21	22	23	24	25	26
21	22	23	24	25	26	27	Tammuz>	27	28	29	30	1	2	3
28	29	30	1	2	3	4	<July	4	5	6	7	8	9	10
5	6	7	8	9	10	11		11	12	13	14	15	16	17
12	13	14	15	16	17	18		18	19	20	21	22	23	24
19	20	21	22	23	24	25		25	26	27	28	29	1	2
26	27	28	29	30	31	1	Ab>	3	4	5	6	7	8	9
2	3	4	5	6	7	8	<August	10	11	12	13	14	15	16
9	10	11	12	13	14	15		17	18	19	20	21	22	23
16	17	18	19	20	21	22		24	25	26	27	28	29	30
23	24	25	26	27	28	29	Elul>	1	2	3	4	5	6	7
30	31	1	2	3	4	5	<Sept	8	9	10	11	12	13	14
6	7	8	9	10	11	12		15	16	17	18	19	20	21
13	14	15	16	17	18	19		22	23	24	25	26	27	28
20	21	22	23	24	25	26	Tishri>	29	1	2	3	4	5	6
27	28	29	30	1	2	3		7	8	9	10	11	12	13
4	5	6	7	8	9	10	<Oct	14	15	16	17	18	19	20
11	12	13	14	15	16	17		21	22	23	24	25	26	27
18	19	20	21	22	23	24		28	29	30	1	2	3	4
25	26	27	28	29	30	31	Heshvan>	5	6	7	8	9	10	11
1	2	3	4	5	6	7	<Nov	12	13	14	15	16	17	18
8	9	10	11	12	13	14		19	20	21	22	23	24	25
15	16	17	18	19	20	21		26	27	28	29	1	2	3
22	23	24	25	26	27	28	Kislev>	4	5	6	7	8	9	10
29	30	1	2	3	4	5	<Dec	11	12	13	14	15	16	17
6	7	8	9	10	11	12		18	19	20	21	22	23	24
13	14	15	16	17	18	19		25	26	27	28	29	1	2
20	21	22	23	24	25	26	Tebeth>	3	4	5	6	7	8	9
27	28	29	30	31				10	11	12	13	14		

AD 538/Dionysius' Cycle I, Year 7/Metonic Year 4/Easter 04 April, Luna 19

S	*M*	*T*	*W*	*T*	*F*	*S*		*S*	*M*	*T*	*W*	*T*	*F*	*S*
					1	2	Tebeth>						15	16
3	4	5	6	7	8	9	<Jan	17	18	19	20	21	22	23
10	11	12	13	14	15	16		24	25	26	27	28	29	1
17	18	19	20	21	22	23	Shebat>	2	3	4	5	6	7	8
24	25	26	27	28	29	30		9	10	11	12	13	14	15
31	1	2	3	4	5	6	<Feb	16	17	18	19	20	21	22
7	8	9	10	11	12	13		23	24	25	26	27	28	29
14	15	16	17	18	19	20	Adar>	30	1	2	3	4	5	6
21	22	23	24	25	26	27		7	8	9	10	11	12	13
28	1	2	3	4	5	6	<March	14	15	16	17	18	19	20
7	8	9	10	11	12	13		21	22	23	24	25	26	27
14	15	16	17	18	19	20		28	29	30	1	2	3	4
21	22	23	24	25	26	27	Nisan>	5	6	7	8	9	10	11
28	29	30	31	1	2	3	<April	12	13	14	15	16	17	18
4	5	6	7	8	9	10		19	20	21	22	23	24	25
11	12	13	14	15	16	17		26	27	28	29	30	1	2
18	19	20	21	22	23	24	Iyar>	3	4	5	6	7	8	9
25	26	27	28	29	30	1		10	11	12	13	14	15	16
2	3	4	5	6	7	8	<May	17	18	19	20	21	22	23
9	10	11	12	13	14	15		24	25	26	27	28	29	1
16	17	18	19	20	21	22	Sivan>	2	3	4	5	6	7	8
23	24	25	26	27	28	29		9	10	11	12	13	14	15
30	31	1	2	3	4	5	<June	16	17	18	19	20	21	22
6	7	8	9	10	11	12		23	24	25	26	27	28	29
13	14	15	16	17	18	19	Tammuz>	30	1	2	3	4	5	6
20	21	22	23	24	25	26		7	8	9	10	11	12	13
27	28	29	30	1	2	3	<July	14	15	16	17	18	19	20
4	5	6	7	8	9	10		21	22	23	24	25	26	27
11	12	13	14	15	16	17	Ab>	28	29	1	2	3	4	5
18	19	20	21	22	23	24		6	7	8	9	10	11	12
25	26	27	28	29	30	31		13	14	15	16	17	18	19
1	2	3	4	5	6	7	<August	20	21	22	23	24	25	26
8	9	10	11	12	13	14		27	28	29	30	1	2	3
15	16	17	18	19	20	21	Elul>	4	5	6	7	8	9	10
22	23	24	25	26	27	28		11	12	13	14	15	16	17
29	30	31	1	2	3	4	<Sept	18	19	20	21	22	23	24
5	6	7	8	9	10	11		25	26	27	28	29	1	2
12	13	14	15	16	17	18	Tishri>	3	4	5	6	7	8	9
19	20	21	22	23	24	25		10	11	12	13	14	15	16
26	27	28	29	30	1	2		17	18	19	20	21	22	23
3	4	5	6	7	8	9	<Oct	24	25	26	27	28	29	30
10	11	12	13	14	15	16	Heshvan>	1	2	3	4	5	6	7
17	18	19	20	21	22	23		8	9	10	11	12	13	14
24	25	26	27	28	29	30		15	16	17	18	19	20	21
31	1	2	3	4	5	6	<Nov	22	23	24	25	26	27	28
7	8	9	10	11	12	13	Kislev>	29	1	2	3	4	5	6
14	15	16	17	18	19	20		7	8	9	10	11	12	13
21	22	23	24	25	26	27		14	15	16	17	18	19	20
28	29	30	1	2	3	4	<Dec	21	22	23	24	25	26	27
5	6	7	8	9	10	11	Tebeth>	28	29	30	1	2	3	4
12	13	14	15	16	17	18		5	6	7	8	9	10	11
19	20	21	22	23	24	25		12	13	14	15	16	17	18
26	27	28	29	30	31			19	20	21	22	23	24	

AD 539/Dionysius' Cycle I, Year 8 (emb.)/Metonic Year 5/Easter 24 April, Luna 20

S	*M*	*T*	*W*	*T*	*F*	*S*		*S*	*M*	*T*	*W*	*T*	*F*	*S*
						1	Tebeth>							25
2	3	4	5	6	7	8	<Jan	26	27	28	29	1	2	3
9	10	11	12	13	14	15	Shevat>	4	5	6	7	8	9	10
16	17	18	19	20	21	22		11	12	13	14	15	16	17
23	24	25	26	27	28	29		18	19	20	21	22	23	24
30	31	1	2	3	4	5	<Feb	25	26	27	28	29	30	1
6	7	8	9	10	11	12	Adar>	2	3	4	5	6	7	8
13	14	15	16	17	18	19		9	10	11	12	13	14	15
20	21	22	23	24	25	26		16	17	18	19	20	21	22
27	28	1	2	3	4	5	<March	23	24	25	26	27	28	29
6	7	8	9	10	11	12	Adar 2>	30	1	2	3	4	5	6
13	14	15	16	17	18	19		7	8	9	10	11	12	13
20	21	22	23	24	25	26		14	15	16	17	18	19	20
27	28	29	30	31	1	2	<April	21	22	23	24	25	26	27
3	4	5	6	7	8	9	Nisan>	28	29	1	2	3	4	5
10	11	12	13	14	15	16		6	7	8	9	10	11	12
17	18	19	20	21	22	23		13	14	15	16	17	18	19
24	25	26	27	28	29	30		20	21	22	23	24	25	26
1	2	3	4	5	6	7	<May	27	28	29	30	1	2	3
8	9	10	11	12	13	14	Iyar>	4	5	6	7	8	9	10
15	16	17	18	19	20	21		11	12	13	14	15	16	17
22	23	24	25	26	27	28		18	19	20	21	22	23	24
29	30	31	1	2	3	4	<June	25	26	27	28	29	1	2
5	6	7	8	9	10	11	Sivan>	3	4	5	6	7	8	9
12	13	14	15	16	17	18		10	11	12	13	14	15	16
19	20	21	22	23	24	25		17	18	19	20	21	22	23
26	27	28	29	30	1	2	<July	24	25	26	27	28	29	30
3	4	5	6	7	8	9	Tammuz>	1	2	3	4	5	6	7
10	11	12	13	14	15	16		8	9	10	11	12	13	14
17	18	19	20	21	22	23		15	16	17	18	19	20	21
24	25	26	27	28	29	30		22	23	24	25	26	27	28
31	1	2	3	4	5	6	<August	29	1	2	3	4	5	6
7	8	9	10	11	12	13	Ab>	7	8	9	10	11	12	13
14	15	16	17	18	19	20		14	15	16	17	18	19	20
21	22	23	24	25	26	27		21	22	23	24	25	26	27
28	29	30	31	1	2	3	<Sept	28	29	30	1	2	3	4
4	5	6	7	8	9	10	Elul>	5	6	7	8	9	10	11
11	12	13	14	15	16	17		12	13	14	15	16	17	18
18	19	20	21	22	23	24		19	20	21	22	23	24	25
25	26	27	28	29	30	1	Tishri	26	27	28	29	1	2	3
2	3	4	5	6	7	8	<Oct	4	5	6	7	8	9	10
9	10	11	12	13	14	15		11	12	13	14	15	16	17
16	17	18	19	20	21	22		18	19	20	21	22	23	24
23	24	25	26	27	28	29	Heshvan>	25	26	27	28	29	30	1
30	31	1	2	3	4	5	<Nov	2	3	4	5	6	7	8
6	7	8	9	10	11	12		9	10	11	12	13	14	15
13	14	15	16	17	18	19		16	17	18	19	20	21	22
20	21	22	23	24	25	26		23	24	25	26	27	28	29
27	28	29	30	1	2	3	Kislev>	1	2	3	4	5	6	7
4	5	6	7	8	9	10	<Dec	8	9	10	11	12	13	14
11	12	13	14	15	16	17		15	16	17	18	19	20	21
18	19	20	21	22	23	24		22	23	24	25	26	27	28
25	26	27	28	29	30	31	Tebeth>	29	30	1	2	3	4	5

AD 540/Dionysius' Cycle I, Year 9/Metonic Year 6/Easter 08 April, Luna 15

S	*M*	*T*	*W*	*T*	*F*	*S*		*S*	*M*	*T*	*W*	*T*	*F*	*S*
1	2	3	4	5	6	7	Tebeth>	6	7	8	9	10	11	12
8	9	10	11	12	13	14	<Jan	13	14	15	16	17	18	19
15	16	17	18	19	20	21		20	21	22	23	24	25	26
22	23	24	25	26	27	28	Shebat>	27	28	29	1	2	3	4
29	30	31	1	2	3	4	<Feb	5	6	7	8	9	10	11
5	6	7	8	9	10	11		12	13	14	15	16	17	18
12	13	14	15	16	17	18		19	20	21	22	23	24	25
19	20	21	22	23	24	25		26	27	28	29	30	1	2
26	27	28	29	1	2	3	Adar>	3	4	5	6	7	8	9
4	5	6	7	8	9	10	<March	10	11	12	13	14	15	16
11	12	13	14	15	16	17		17	18	19	20	21	22	23
18	19	20	21	22	23	24		24	25	26	27	28	29	30
25	26	27	28	29	30	31	Nisan>	1	2	3	4	5	6	7
1	2	3	4	5	6	7	<April	8	9	10	11	12	13	14
8	9	10	11	12	13	14		15	16	17	18	19	20	21
15	16	17	18	19	20	21		22	23	24	25	26	27	28
22	23	24	25	26	27	28	Iyar>	29	30	1	2	3	4	5
29	30	1	2	3	4	5	<May	6	7	8	9	10	11	12
6	7	8	9	10	11	12		13	14	15	16	17	18	19
13	14	15	16	17	18	19		20	21	22	23	24	25	26
20	21	22	23	24	25	26		27	28	29	1	2	3	4
27	28	29	30	31	1	2	Sivan>	5	6	7	8	9	10	11
3	4	5	6	7	8	9	<June	12	13	14	15	16	17	18
10	11	12	13	14	15	16		19	20	21	22	23	24	25
17	18	19	20	21	22	23		26	27	28	29	30	1	2
24	25	26	27	28	29	30	Tammuz>	3	4	5	6	7	8	9
1	2	3	4	5	6	7	<July	10	11	12	13	14	15	16
8	9	10	11	12	13	14		17	18	19	20	21	22	23
15	16	17	18	19	20	21		24	25	26	27	28	29	1
22	23	24	25	26	27	28	Ab>	2	3	4	5	6	7	8
29	30	31	1	2	3	4	<August	9	10	11	12	13	14	15
5	6	7	8	9	10	11		16	17	18	19	20	21	22
12	13	14	15	16	17	18		23	24	25	26	27	28	29
19	20	21	22	23	24	25	Elul>	30	1	2	3	4	5	6
26	27	28	29	30	31	1		7	8	9	10	11	12	13
2	3	4	5	6	7	8	<Sept	14	15	16	17	18	19	20
9	10	11	12	13	14	15		21	22	23	24	25	26	27
16	17	18	19	20	21	22		28	29	1	2	3	4	5
23	24	25	26	27	28	29	Tishri>	6	7	8	9	10	11	12
30	1	2	3	4	5	6	<Oct	13	14	15	16	17	18	19
7	8	9	10	11	12	13		20	21	22	23	24	25	26
14	15	16	17	18	19	20		27	28	29	30	1	2	3
21	22	23	24	25	26	27	Heshvan>	4	5	6	7	8	9	10
28	29	30	31	1	2	3		11	12	13	14	15	16	17
4	5	6	7	8	9	10	<Nov	18	19	20	21	22	23	24
11	12	13	14	15	16	17		25	26	27	28	29	1	2
18	19	20	21	22	23	24	Kislev>	3	4	5	6	7	8	9
25	26	27	28	29	30	1		10	11	12	13	14	15	16
2	3	4	5	6	7	8	<Dec	17	18	19	20	21	22	23
9	10	11	12	13	14	15		24	25	26	27	28	29	1
16	17	18	19	20	21	22	Tebeth>	2	3	4	5	6	7	8
23	24	25	26	27	28	29		9	10	11	12	13	14	15
30	31							16	17					

AD 541/Dionysius' Cycle I, Year 10/Metonic Year 7/Easter 31 March, Luna 18

S	*M*	*T*	*W*	*T*	*F*	*S*		*S*	*M*	*T*	*W*	*T*	*F*	*S*
		1	2	3	4	5	Tebeth>			18	19	20	21	22
6	7	8	9	10	11	12	<Jan	23	24	25	26	27	28	29
13	14	15	16	17	18	19		1	2	3	4	5	6	7
20	21	22	23	24	25	26		8	9	10	11	12	13	14
27	28	29	30	31	1	2	<Feb	15	16	17	18	19	20	21
3	4	5	6	7	8	9		22	23	24	25	26	27	28
10	11	12	13	14	15	16		29	30	1	2	3	4	5
17	18	19	20	21	22	23	Adar>	6	7	8	9	10	11	12
24	25	26	27	28	1	2		13	14	15	16	17	18	19
3	4	5	6	7	8	9	<March	20	21	22	23	24	25	26
10	11	12	13	14	15	16		27	28	29	30	1	2	3
17	18	19	20	21	22	23	Nisan>	4	5	6	7	8	9	10
24	25	26	27	28	29	30		11	12	13	14	15	16	17
31	1	2	3	4	5	6	<April	18	19	20	21	22	23	24
7	8	9	10	11	12	13		25	26	27	28	29	30	1
14	15	16	17	18	19	20	Iyar>	2	3	4	5	6	7	8
21	22	23	24	25	26	27		9	10	11	12	13	14	15
28	29	30	1	2	3	4	<May	16	17	18	19	20	21	22
5	6	7	8	9	10	11		23	24	25	26	27	28	29
12	13	14	15	16	17	18	Sivan>	1	2	3	4	5	6	7
19	20	21	22	23	24	25		8	9	10	11	12	13	14
26	27	28	29	30	31	1		15	16	17	18	19	20	21
2	3	4	5	6	7	8	<June	22	23	24	25	26	27	28
9	10	11	12	13	14	15		29	30	1	2	3	4	5
16	17	18	19	20	21	22	Tammuz>	6	7	8	9	10	11	12
23	24	25	26	27	28	29		13	14	15	16	17	18	19
30	1	2	3	4	5	6	<July	20	21	22	23	24	25	26
7	8	9	10	11	12	13	Ab>	27	28	29	1	2	3	4
14	15	16	17	18	19	20		5	6	7	8	9	10	11
21	22	23	24	25	26	27		12	13	14	15	16	17	18
28	29	30	31	1	2	3		19	20	21	22	23	24	25
4	5	6	7	8	9	10	<August	26	27	28	29	30	1	2
11	12	13	14	15	16	17	Elul>	3	4	5	6	7	8	9
18	19	20	21	22	23	24		10	11	12	13	14	15	16
25	26	27	28	29	30	31		17	18	19	20	21	22	23
1	2	3	4	5	6	7	<Sept	24	25	26	27	28	29	1
8	9	10	11	12	13	14	Tishri	2	3	4	5	6	7	8
15	16	17	18	19	20	21		9	10	11	12	13	14	15
22	23	24	25	26	27	28		16	17	18	19	20	21	22
29	30	1	2	3	4	5	<Oct	23	24	25	26	27	28	29
6	7	8	9	10	11	12	Heshvan>	30	1	2	3	4	5	6
13	14	15	16	17	18	19		7	8	9	10	11	12	13
20	21	22	23	24	25	26		14	15	16	17	18	19	20
27	28	29	30	31	1	2		21	22	23	24	25	26	27
3	4	5	6	7	8	9	<Nov	28	29	1	2	3	4	5
10	11	12	13	14	15	16	Kislev>	6	7	8	9	10	11	12
17	18	19	20	21	22	23		13	14	15	16	17	18	19
24	25	26	27	28	29	30		20	21	22	23	24	25	26
1	2	3	4	5	6	7	<Dec	27	28	29	30	1	2	3
8	9	10	11	12	13	14	Tebeth>	4	5	6	7	8	9	10
15	16	17	18	19	20	21		11	12	13	14	15	16	17
22	23	24	25	26	27	28		18	19	20	21	22	23	24
29	30	31						25	26	27				

AD 542/Dionysius' Cycle I, Year 11 (emb.)/Metonic Year 8/Easter 20 April, Luna 19

S	*M*	*T*	*W*	*T*	*F*	*S*		*S*	*M*	*T*	*W*	*T*	*F*	*S*
			1	2	3	4	Tebeth>				28	29	1	2
5	6	7	8	9	10	11	<Jan	3	4	5	6	7	8	9
12	13	14	15	16	17	18	Shevat>	10	11	12	13	14	15	16
19	20	21	22	23	24	25		17	18	19	20	21	22	23
26	27	28	29	30	31	1		24	25	26	27	28	29	30
2	3	4	5	6	7	8	<Feb	1	2	3	4	5	6	7
9	10	11	12	13	14	15	Adar>	8	9	10	11	12	13	14
16	17	18	19	20	21	22		15	16	17	18	19	20	21
23	24	25	26	27	28	1	<March	22	23	24	25	26	27	28
2	3	4	5	6	7	8	Adar 2>	29	30	1	2	3	4	5
9	10	11	12	13	14	15		6	7	8	9	10	11	12
16	17	18	19	20	21	22		13	14	15	16	17	18	19
23	24	25	26	27	28	29		20	21	22	23	24	25	26
30	31	1	2	3	4	5	<April	27	28	29	1	2	3	4
6	7	8	9	10	11	12	Nisan>	5	6	7	8	9	10	11
13	14	15	16	17	18	19		12	13	14	15	16	17	18
20	21	22	23	24	25	26		19	20	21	22	23	24	25
27	28	29	30	1	2	3	<May	26	27	28	29	30	1	2
4	5	6	7	8	9	10	Iyar>	3	4	5	6	7	8	9
11	12	13	14	15	16	17		10	11	12	13	14	15	16
18	19	20	21	22	23	24		17	18	19	20	21	22	23
25	26	27	28	29	30	31	Sivan>	24	25	26	27	28	29	1
1	2	3	4	5	6	7	<June	2	3	4	5	6	7	8
8	9	10	11	12	13	14		9	10	11	12	13	14	15
15	16	17	18	19	20	21		16	17	18	19	20	21	22
22	23	24	25	26	27	28		23	24	25	26	27	28	29
29	30	1	2	3	4	5	<July	30	1	2	3	4	5	6
6	7	8	9	10	11	12	Tamuuz>	7	8	9	10	11	12	13
13	14	15	16	17	18	19		14	15	16	17	18	19	20
20	21	22	23	24	25	26		21	22	23	24	25	26	27
27	28	29	30	31	1	2	Ab>	28	29	1	2	3	4	5
3	4	5	6	7	8	9	<August	6	7	8	9	10	11	12
10	11	12	13	14	15	16		13	14	15	16	17	18	19
17	18	19	20	21	22	23		20	21	22	23	24	25	26
24	25	26	27	28	29	30	Elul>	27	28	29	30	1	2	3
31	1	2	3	4	5	6	<Sept	4	5	6	7	8	9	10
7	8	9	10	11	12	13		11	12	13	14	15	16	17
14	15	16	17	18	19	20		18	19	20	21	22	23	24
21	22	23	24	25	26	27	Tishri>	25	26	27	28	29	1	2
28	29	30	1	2	3	4	<Oct	3	4	5	6	7	8	9
5	6	7	8	9	10	11		10	11	12	13	14	15	16
12	13	14	15	16	17	18		17	18	19	20	21	22	23
19	20	21	22	23	24	25		24	25	26	27	28	29	30
26	27	28	29	30	31	1	Heshvan>	1	2	3	4	5	6	7
2	3	4	5	6	7	8	<Nov	8	9	10	11	12	13	14
9	10	11	12	13	14	15		15	16	17	18	19	20	21
16	17	18	19	20	21	22		22	23	24	25	26	27	28
23	24	25	26	27	28	29	Kislev>	29	1	2	3	4	5	6
30	1	2	3	4	5	6	<Dec	7	8	9	10	11	12	13
7	8	9	10	11	12	13		14	15	16	17	18	19	20
14	15	16	17	18	19	20		21	22	23	24	25	26	27
21	22	23	24	25	26	27	Tebeth>	28	29	1	2	3	4	5
28	29	30	31					6	7	8	9			

AD 543/Dionysius' Cycle I, Year 12/Metonic Year 8/Easter 05 April, Luna 15

S	*M*	*T*	*W*	*T*	*F*	*S*		*S*	*M*	*T*	*W*	*T*	*F*	*S*
				1	2	3	Tebeth>					10	11	12
4	5	6	7	8	9	10	<Jan	13	14	15	16	17	18	19
11	12	13	14	15	16	17		20	21	22	23	24	25	26
18	19	20	21	22	23	24		27	28	29	1	2	3	4
25	26	27	28	29	30	31	Shevat>	5	6	7	8	9	10	11
1	2	3	4	5	6	7	<Feb	12	13	14	15	16	17	18
8	9	10	11	12	13	14		19	20	21	22	23	24	25
15	16	17	18	19	20	21		26	27	28	29	30	1	2
22	23	24	25	26	27	28	Adar>	3	4	5	6	7	8	9
1	2	3	4	5	6	7	<March	10	11	12	13	14	15	16
8	9	10	11	12	13	14		17	18	19	20	21	22	23
15	16	17	18	19	20	21		24	25	26	27	28	29	30
22	23	24	25	26	27	28	Nisan>	1	2	3	4	5	6	7
29	30	31	1	2	3	4	<April	8	9	10	11	12	13	14
5	6	7	8	9	10	11		15	16	17	18	19	20	21
12	13	14	15	16	17	18		22	23	24	25	26	27	28
19	20	21	22	23	24	25		29	30	1	2	3	4	5
26	27	28	29	30	1	2	Iyar>	6	7	8	9	10	11	12
3	4	5	6	7	8	9	<May	13	14	15	16	17	18	19
10	11	12	13	14	15	16		20	21	22	23	24	25	26
17	18	19	20	21	22	23		27	28	29	1	2	3	4
24	25	26	27	28	29	30	Sivan>	5	6	7	8	9	10	11
31	1	2	3	4	5	6	<June	12	13	14	15	16	17	18
7	8	9	10	11	12	13		19	20	21	22	23	24	25
14	15	16	17	18	19	20		26	27	28	29	30	1	2
21	22	23	24	25	26	27	Tammuz>	3	4	5	6	7	8	9
28	29	30	1	2	3	4	<July	10	11	12	13	14	15	16
5	6	7	8	9	10	11		17	18	19	20	21	22	23
12	13	14	15	16	17	18		24	25	26	27	28	29	1
19	20	21	22	23	24	25	Ab>	2	3	4	5	6	7	8
26	27	28	29	30	31	1		9	10	11	12	13	14	15
2	3	4	5	6	7	8	<August	16	17	18	19	20	21	22
9	10	11	12	13	14	15		23	24	25	26	27	28	29
16	17	18	19	20	21	22		30	1	2	3	4	5	6
23	24	25	26	27	28	29	Elul>	7	8	9	10	11	12	13
30	31	1	2	3	4	5	<Sept	14	15	16	17	18	19	20
6	7	8	9	10	11	12		21	22	23	24	25	26	27
13	14	15	16	17	18	19		28	29	1	2	3	4	5
20	21	22	23	24	25	26	Tishri>	6	7	8	9	10	11	12
27	28	29	30	1	2	3		13	14	15	16	17	18	19
4	5	6	7	8	9	10	<Oct	20	21	22	23	24	25	26
11	12	13	14	15	16	17		27	28	29	30	1	2	3
18	19	20	21	22	23	24	Heshvan>	4	5	6	7	8	9	10
25	26	27	28	29	30	31		11	12	13	14	15	16	17
1	2	3	4	5	6	7	<Nov	18	19	20	21	22	23	24
8	9	10	11	12	13	14		25	26	27	28	29	30	1
15	16	17	18	19	20	21	Kislev>	2	3	4	5	6	7	8
22	23	24	25	26	27	28		9	10	11	12	13	14	15
29	30	1	2	3	4	5	<Dec	16	17	18	19	20	21	22
6	7	8	9	10	11	12		23	24	25	26	27	28	29
13	14	15	16	17	18	19		30	1	2	3	4	5	6
20	21	22	23	24	25	26	Tebeth>	7	8	9	10	11	12	13
27	28	29	30	31				14	15	16	17	18		

AD 544/Dionysius' Cycle I, Year 13/Metonic Year 10/Easter 27 March, Luna 17

S	*M*	*T*	*W*	*T*	*F*	*S*		*S*	*M*	*T*	*W*	*T*	*F*	*S*
					1	2	Tebeth>						19	20
3	4	5	6	7	8	9	<Jan	21	22	23	24	25	26	27
10	11	12	13	14	15	16		28	29	1	2	3	4	5
17	18	19	20	21	22	23	Shevat>	6	7	8	9	10	11	12
24	25	26	27	28	29	30		13	14	15	16	17	18	19
31	1	2	3	4	5	6	<Feb	20	21	22	23	24	25	26
7	8	9	10	11	12	13		27	28	29	30	1	2	3
14	15	16	17	18	19	20	Adar>	4	5	6	7	8	9	10
21	22	23	24	25	26	27		11	12	13	14	15	16	17
28	29	1	2	3	4	5		18	19	20	21	22	23	24
6	7	8	9	10	11	12	<March	25	26	27	28	29	1	2
13	14	15	16	17	18	19	Nisan>	3	4	5	6	7	8	9
20	21	22	23	24	25	26		10	11	12	13	14	15	16
27	28	29	30	31	1	2		17	18	19	20	21	22	23
3	4	5	6	7	8	9	<April	24	25	26	27	28	29	30
10	11	12	13	14	15	16	Iyar>	1	2	3	4	5	6	7
17	18	19	20	21	22	23		8	9	10	11	12	13	14
24	25	26	27	28	29	30		15	16	17	18	19	20	21
1	2	3	4	5	6	7	<May	22	23	24	25	26	27	28
8	9	10	11	12	13	14		29	1	2	3	4	5	6
15	16	17	18	19	20	21	Sivan>	7	8	9	10	11	12	13
22	23	24	25	26	27	28		14	15	16	17	18	19	20
29	30	31	1	2	3	4		21	22	23	24	25	26	27
5	6	7	8	9	10	11	<June	28	29	30	1	2	3	4
12	13	14	15	16	17	18	Tammuz>	5	6	7	8	9	10	11
19	20	21	22	23	24	25		12	13	14	15	16	17	18
26	27	28	29	30	1	2		19	20	21	22	23	24	25
3	4	5	6	7	8	9	<July	26	27	28	29	1	2	3
10	11	12	13	14	15	16	Ab>	4	5	6	7	8	9	10
17	18	19	20	21	22	23		11	12	13	14	15	16	17
24	25	26	27	28	29	30		18	19	20	21	22	23	24
31	1	2	3	4	5	6	<August	25	26	27	28	29	30	1
7	8	9	10	11	12	13	Elul>	2	3	4	5	6	7	8
14	15	16	17	18	19	20		9	10	11	12	13	14	15
21	22	23	24	25	26	27		16	17	18	19	20	21	22
28	29	30	31	1	2	3		23	24	25	26	27	28	29
4	5	6	7	8	9	10	<Sept	1	2	3	4	5	6	7
11	12	13	14	15	16	17	Tishri>	8	9	10	11	12	13	14
18	19	20	21	22	23	24		15	16	17	18	19	20	21
25	26	27	28	29	30	1		22	23	24	25	26	27	28
2	3	4	5	6	7	8	<Oct	29	30	1	2	3	4	5
9	10	11	12	13	14	15	Heshvan>	6	7	8	9	10	11	12
16	17	18	19	20	21	22		13	14	15	16	17	18	19
23	24	25	26	27	28	29		20	21	22	23	24	25	26
30	31	1	2	3	4	5	<Nov	27	28	29	1	2	3	4
6	7	8	9	10	11	12	Kislev>	5	6	7	8	9	10	11
13	14	15	16	17	18	19		12	13	14	15	16	17	18
20	21	22	23	24	25	26		19	20	21	22	23	24	25
27	28	29	30	1	2	3	Tebeth>	26	27	28	29	30	1	2
4	5	6	7	8	9	10	<Dec	3	4	5	6	7	8	9
11	12	13	14	15	16	17		10	11	12	13	14	15	16
18	19	20	21	22	23	24		17	18	19	20	21	22	23
25	26	27	28	29	30	31	Shevat>	24	25	26	27	28	29	1

AD 545/Dionysius' Cycle I, Year 14 (emb.)/Metonic Year 11/Easter 16 April, Luna 18

S	*M*	*T*	*W*	*T*	*F*	*S*		*S*	*M*	*T*	*W*	*T*	*F*	*S*
1	2	3	4	5	6	7	Shevat>	2	3	4	5	6	7	8
8	9	10	11	12	13	14	<Jan	9	10	11	12	13	14	15
15	16	17	18	19	20	21		16	17	18	19	20	21	22
22	23	24	25	26	27	28		23	24	25	26	27	28	29
29	30	31	1	2	3	4	Adar>	30	1	2	3	4	5	6
5	6	7	8	9	10	11	<Feb	7	8	9	10	11	12	13
12	13	14	15	16	17	18		14	15	16	17	18	19	20
19	20	21	22	23	24	25		21	22	23	24	25	26	27
26	27	28	1	2	3	4	Adar 2>	28	29	30	1	2	3	4
5	6	7	8	9	10	11	<March	5	6	7	8	9	10	11
12	13	14	15	16	17	18		12	13	14	15	16	17	18
19	20	21	22	23	24	25		19	20	21	22	23	24	25
26	27	28	29	30	31	1	Nisan>	26	27	28	29	1	2	3
2	3	4	5	6	7	8	<April	4	5	6	7	8	9	10
9	10	11	12	13	14	15		11	12	13	14	15	16	17
16	17	18	19	20	21	22		18	19	20	21	22	23	24
23	24	25	26	27	28	29		25	26	27	28	29	30	1
30	1	2	3	4	5	6	Iyar>	2	3	4	5	6	7	8
7	8	9	10	11	12	13	<May	9	10	11	12	13	14	15
14	15	16	17	18	19	20		16	17	18	19	20	21	22
21	22	23	24	25	26	27		23	24	25	26	27	28	29
28	29	30	31	1	2	3	Sivan>	1	2	3	4	5	6	7
4	5	6	7	8	9	10	<June	8	9	10	11	12	13	14
11	12	13	14	15	16	17		15	16	17	18	19	20	21
18	19	20	21	22	23	24		22	23	24	25	26	27	28
25	26	27	28	29	30	1	Tammuz>	29	30	1	2	3	4	5
2	3	4	5	6	7	8	<July	6	7	8	9	10	11	12
9	10	11	12	13	14	15		13	14	15	16	17	18	19
16	17	18	19	20	21	22		20	21	22	23	24	25	26
23	24	25	26	27	28	29	Ab>	27	28	29	1	2	3	4
30	31	1	2	3	4	5	<August	5	6	7	8	9	10	11
6	7	8	9	10	11	12		12	13	14	15	16	17	18
13	14	15	16	17	18	19		19	20	21	22	23	24	25
20	21	22	23	24	25	26		26	27	28	29	30	1	2
27	28	29	30	31	1	2	Elul>	3	4	5	6	7	8	9
3	4	5	6	7	8	9	<Sept	10	11	12	13	14	15	16
10	11	12	13	14	15	16		17	18	19	20	21	22	23
17	18	19	20	21	22	23		24	25	26	27	28	29	1
24	25	26	27	28	29	30	Tishri>	2	3	4	5	6	7	8
1	2	3	4	5	6	7	<Oct	9	10	11	12	13	14	15
8	9	10	11	12	13	14		16	17	18	19	20	21	22
15	16	17	18	19	20	21		23	24	25	26	27	28	29
22	23	24	25	26	27	28	Heshvan>	30	1	2	3	4	5	6
29	30	31	1	2	3	4		7	8	9	10	11	12	13
5	6	7	8	9	10	11	<Nov	14	15	16	17	18	19	20
12	13	14	15	16	17	18		21	22	23	24	25	26	27
19	20	21	22	23	24	25		28	29	1	2	3	4	5
26	27	28	29	30	1	2	Kislev>	6	7	8	9	10	11	12
3	4	5	6	7	8	9	<Dec	13	14	15	16	17	18	19
10	11	12	13	14	15	16		20	21	22	23	24	25	26
17	18	19	20	21	22	23		27	28	29	1	2	3	4
24	25	26	27	28	29	30	Tebeth>	5	6	7	8	9	10	11
31								12						

AD 546/Dionysius' Cycle I, Year 15/Metonic Year 12/Easter 08 April, Luna 21

S	*M*	*T*	*W*	*T*	*F*	*S*		*S*	*M*	*T*	*W*	*T*	*F*	*S*
	1	2	3	4	5	6	Tebeth>		13	14	15	16	17	18
7	8	9	10	11	12	13	<Jan	19	20	21	22	23	24	25
14	15	16	17	18	19	20		26	27	28	29	1	2	3
21	22	23	24	25	26	27	Shevat>	4	5	6	7	8	9	10
28	29	30	31	1	2	3		11	12	13	14	15	16	17
4	5	6	7	8	9	10	<Feb	18	19	20	21	22	23	24
11	12	13	14	15	16	17		25	26	27	28	29	30	1
18	19	20	21	22	23	24	Adar>	2	3	4	5	6	7	8
25	26	27	28	1	2	3		9	10	11	12	13	14	15
4	5	6	7	8	9	10	<March	16	17	18	19	20	21	22
11	12	13	14	15	16	17		23	24	25	26	27	28	29
18	19	20	21	22	23	24	Nisan>	30	1	2	3	4	5	6
25	26	27	28	29	30	31		7	8	9	10	11	12	13
1	2	3	4	5	6	7	<April	14	15	16	17	18	19	20
8	9	10	11	12	13	14		21	22	23	24	25	26	27
15	16	17	18	19	20	21		28	29	30	1	2	3	4
22	23	24	25	26	27	28	Iyar>	5	6	7	8	9	10	11
29	30	1	2	3	4	5		12	13	14	15	16	17	18
6	7	8	9	10	11	12	<May	19	20	21	22	23	24	25
13	14	15	16	17	18	19		26	27	28	29	1	2	3
20	21	22	23	24	25	26	Sivan>	4	5	6	7	8	9	10
27	28	29	30	31	1	2		11	12	13	14	15	16	17
3	4	5	6	7	8	9	<June	18	19	20	21	22	23	24
10	11	12	13	14	15	16		25	26	27	28	29	30	1
17	18	19	20	21	22	23	Tammuz>	2	3	4	5	6	7	8
24	25	26	27	28	29	30		9	10	11	12	13	14	15
1	2	3	4	5	6	7	<July	16	17	18	19	20	21	22
8	9	10	11	12	13	14		23	24	25	26	27	28	29
15	16	17	18	19	20	21	Ab>	1	2	3	4	5	6	7
22	23	24	25	26	27	28		8	9	10	11	12	13	14
29	30	31	1	2	3	4		15	16	17	18	19	20	21
5	6	7	8	9	10	11	<August	22	23	24	25	26	27	28
12	13	14	15	16	17	18		29	30	1	2	3	4	5
19	20	21	22	23	24	25	Elul>	6	7	8	9	10	11	12
26	27	28	29	30	31	1		13	14	15	16	17	18	19
2	3	4	5	6	7	8	<Sept	20	21	22	23	24	25	26
9	10	11	12	13	14	15		27	28	29	1	2	3	4
16	17	18	19	20	21	22	Tishri>	5	6	7	8	9	10	11
23	24	25	26	27	28	29		12	13	14	15	16	17	18
30	1	2	3	4	5	6	<Oct	19	20	21	22	23	24	25
7	8	9	10	11	12	13		26	27	28	29	30	1	2
14	15	16	17	18	19	20	Heshvan>	3	4	5	6	7	8	9
21	22	23	24	25	26	27		10	11	12	13	14	15	16
28	29	30	31	1	2	3		17	18	19	20	21	22	23
4	5	6	7	8	9	10	<Nov	24	25	26	27	28	29	1
11	12	13	14	15	16	17	Kislev>	2	3	4	5	6	7	8
18	19	20	21	22	23	24		9	10	11	12	13	14	15
25	26	27	28	29	30	1		16	17	18	19	20	21	22
2	3	4	5	6	7	8	<Dec	23	24	25	26	27	28	29
9	10	11	12	13	14	15	Tebeth>	1	2	3	4	5	6	7
16	17	18	19	20	21	22		8	9	10	11	12	13	14
23	24	25	26	27	28	29		15	16	17	18	19	20	21
30	31							22	23					

AD 547/Dionysius' Cycle I, Year 16/Metonic Year 13/Easter 24 March, Luna 17

S	*M*	*T*	*W*	*T*	*F*	*S*		*S*	*M*	*T*	*W*	*T*	*F*	*S*
		1	2	3	4	5	Tebeth>			24	25	26	27	28
6	7	8	9	10	11	12	<Jan	29	1	2	3	4	5	6
13	14	15	16	17	18	19	Shevat>	7	8	9	10	11	12	13
20	21	22	23	24	25	26		14	15	16	17	18	19	20
27	28	29	30	31	1	2		21	22	23	24	25	26	27
3	4	5	6	7	8	9	<Feb	28	29	30	1	2	3	4
10	11	12	13	14	15	16	Adar>	5	6	7	8	9	10	11
17	18	19	20	21	22	23		12	13	14	15	16	17	18
24	25	26	27	28	1	2		19	20	21	22	23	24	25
3	4	5	6	7	8	9	<March	26	27	28	29	30	1	2
10	11	12	13	14	15	16	Nisan>	3	4	5	6	7	8	9
17	18	19	20	21	22	23		10	11	12	13	14	15	16
24	25	26	27	28	29	30		17	18	19	20	21	22	23
31	1	2	3	4	5	6	<April	24	25	26	27	28	29	30
7	8	9	10	11	12	13	Iyar>	1	2	3	4	5	6	7
14	15	16	17	18	19	20		8	9	10	11	12	13	14
21	22	23	24	25	26	27		15	16	17	18	19	20	21
28	29	30	1	2	3	4	<May	22	23	24	25	26	27	28
5	6	7	8	9	10	11		29	1	2	3	4	5	6
12	13	14	15	16	17	18	Sivan>	7	8	9	10	11	12	13
19	20	21	22	23	24	25		14	15	16	17	18	19	20
26	27	28	29	30	31	1		21	22	23	24	25	26	27
2	3	4	5	6	7	8	<June	28	29	30	1	2	3	4
9	10	11	12	13	14	15	Tammuz>	5	6	7	8	9	10	11
16	17	18	19	20	21	22		12	13	14	15	16	17	18
23	24	25	26	27	28	29		19	20	21	22	23	24	25
30	1	2	3	4	5	6	<July	26	27	28	29	1	2	3
7	8	9	10	11	12	13	Ab>	4	5	6	7	8	9	10
14	15	16	17	18	19	20		11	12	13	14	15	16	17
21	22	23	24	25	26	27		18	19	20	21	22	23	24
28	29	30	31	1	2	3	<August	25	26	27	28	29	30	1
4	5	6	7	8	9	10	Elul>	2	3	4	5	6	7	8
11	12	13	14	15	16	17		9	10	11	12	13	14	15
18	19	20	21	22	23	24		16	17	18	19	20	21	22
25	26	27	28	29	30	31		23	24	25	26	27	28	29
1	2	3	4	5	6	7	<Sept	1	2	3	4	5	6	7
8	9	10	11	12	13	14	Tishri>	8	9	10	11	12	13	14
15	16	17	18	19	20	21		15	16	17	18	19	20	21
22	23	24	25	26	27	28		22	23	24	25	26	27	28
29	30	1	2	3	4	5	<Oct	29	30	1	2	3	4	5
6	7	8	9	10	11	12	Heshvan>	6	7	8	9	10	11	12
13	14	15	16	17	18	19		13	14	15	16	17	18	19
20	21	22	23	24	25	26		20	21	22	23	24	25	26
27	28	29	30	31	1	2	Kislev>	27	28	29	30	1	2	3
3	4	5	6	7	8	9	<Nov	4	5	6	7	8	9	10
10	11	12	13	14	15	16		11	12	13	14	15	16	17
17	18	19	20	21	22	23		18	19	20	21	22	23	24
24	25	26	27	28	29	30	Tebeth>	25	26	27	28	29	30	1
1	2	3	4	5	6	7	<Dec	2	3	4	5	6	7	8
8	9	10	11	12	13	14		9	10	11	12	13	14	15
15	16	17	18	19	20	21		16	17	18	19	20	21	22
22	23	24	25	26	27	28		23	24	25	26	27	28	29
29	30	31					Shevat>	1	2	3				

AD 548/Dionysius' Cycle I, Year 17 (emb.)/Metonic Year 14/Easter 12 April, Luna 17

S	*M*	*T*	*W*	*T*	*F*	*S*		*S*	*M*	*T*	*W*	*T*	*F*	*S*
			1	2	3	4	Shevat>				4	5	6	7
5	6	7	8	9	10	11	<Jan	8	9	10	11	12	13	14
12	13	14	15	16	17	18		15	16	17	18	19	20	21
19	20	21	22	23	24	25		22	23	24	25	26	27	28
26	27	28	29	30	31	1	Adar>	29	30	1	2	3	4	5
2	3	4	5	6	7	8	<Feb	6	7	8	9	10	11	12
9	10	11	12	13	14	15		13	14	15	16	17	18	19
16	17	18	19	20	21	22		20	21	22	23	24	25	26
23	24	25	26	27	28	29		27	28	29	30	1	2	3
1	2	3	4	5	6	7	Adar 2>	4	5	6	7	8	9	10
8	9	10	11	12	13	14	<March	11	12	13	14	15	16	17
15	16	17	18	19	20	21		18	19	20	21	22	23	24
22	23	24	25	26	27	28		25	26	27	28	29	1	2
29	30	31	1	2	3	4	Nisan>	3	4	5	6	7	8	9
5	6	7	8	9	10	11	<April	10	11	12	13	14	15	16
12	13	14	15	16	17	18		17	18	19	20	21	22	23
19	20	21	22	23	24	25		24	25	26	27	28	29	30
26	27	28	29	30	1	2	Iyar>	1	2	3	4	5	6	7
3	4	5	6	7	8	9	<May	8	9	10	11	12	13	14
10	11	12	13	14	15	16		15	16	17	18	19	20	21
17	18	19	20	21	22	23		22	23	24	25	26	27	28
24	25	26	27	28	29	30	Sivan>	29	1	2	3	4	5	6
31	1	2	3	4	5	6	<June	7	8	9	10	11	12	13
7	8	9	10	11	12	13		14	15	16	17	18	19	20
14	15	16	17	18	19	20		21	22	23	24	25	26	27
21	22	23	24	25	26	27		28	29	30	1	2	3	4
28	29	30	1	2	3	4	Tammuz>	5	6	7	8	9	10	11
5	6	7	8	9	10	11	<July	12	13	14	15	16	17	18
12	13	14	15	16	17	18		19	20	21	22	23	24	25
19	20	21	22	23	24	25		26	27	28	29	1	2	3
26	27	28	29	30	31	1	Ab>	4	5	6	7	8	9	10
2	3	4	5	6	7	8	<August	11	12	13	14	15	16	17
9	10	11	12	13	14	15		18	19	20	21	22	23	24
16	17	18	19	20	21	22		25	26	27	28	29	30	1
23	24	25	26	27	28	29	Elul>	2	3	4	5	6	7	8
30	31	1	2	3	4	5	<Sept	9	10	11	12	13	14	15
6	7	8	9	10	11	12		16	17	18	19	20	21	22
13	14	15	126	17	18	19		23	24	25	26	27	28	29
20	21	22	23	24	25	26	Tishri>	1	2	3	4	5	6	7
27	28	29	30	1	2	3		8	9	10	11	12	13	14
4	5	6	7	8	9	10	<Oct	15	16	17	18	19	20	21
11	12	13	14	15	16	17		22	23	24	25	26	27	28
18	19	20	21	22	23	24		29	30	1	2	3	4	5
25	26	27	28	29	30	31	Heshvan>	6	7	8	9	10	11	12
1	2	3	4	5	6	7	<Nov	13	14	15	16	17	18	19
8	9	10	11	12	13	14		20	21	22	23	24	25	26
15	16	17	18	19	20	21		27	28	29	1	2	3	4
22	23	24	25	26	27	28	Kislev>	5	6	7	8	9	10	11
29	30	1	2	3	4	5		12	13	14	15	16	17	18
6	7	8	9	10	11	12	<Dec	19	20	21	22	23	24	25
13	14	15	16	17	18	19		26	27	28	29	1	2	3
20	21	22	23	24	25	26	Tebeth>	4	5	6	7	8	9	10
27	28	29	30	31				11	12	13	14	15		

AD 549/Dionysius' Cycle I, Year 18/Metonic Year 15/Easter 04 April, Luna 20

S	*M*	*T*	*W*	*T*	*F*	*S*		*S*	*M*	*T*	*W*	*T*	*F*	*S*
					1	2	Tebeth>						16	17
3	4	5	6	7	8	9	<Jan	18	19	20	21	22	23	24
10	11	12	13	14	15	16		25	26	27	28	29	1	2
17	18	19	20	21	22	23	Shevat>	3	4	5	6	7	8	9
24	25	26	27	28	29	30		10	11	12	13	14	15	16
31	1	2	3	4	5	6	<Feb	17	18	19	20	21	22	23
7	8	9	10	11	12	13		24	25	26	27	28	29	30
14	15	16	17	18	19	20	Adar>	1	2	3	4	5	6	7
21	22	23	24	25	26	27		8	9	10	11	12	13	14
28	1	2	3	4	5	6	<March	15	16	17	18	19	20	21
7	8	9	10	11	12	13		22	23	24	25	26	27	28
14	15	16	17	18	19	20		29	30	1	2	3	4	5
21	22	23	24	25	26	27	Nisan>	6	7	8	9	10	11	12
28	29	30	31	1	2	3	<April	13	14	15	16	17	18	19
4	5	6	7	8	9	10		20	21	22	23	24	25	26
11	12	13	14	15	16	17		27	28	29	30	1	2	3
18	19	20	21	22	23	24	Iyar>	4	5	6	7	8	9	10
25	26	27	28	29	30	1		11	12	13	14	15	16	17
2	3	4	5	6	7	8	<May	18	19	20	21	22	23	24
9	10	11	12	13	14	15		25	26	27	28	29	1	2
16	17	18	19	20	21	22	Sivan>	3	4	5	6	7	8	9
23	24	25	26	27	28	29		10	11	12	13	14	15	16
30	31	1	2	3	4	5	<June	17	18	19	20	21	22	23
6	7	8	9	10	11	12		24	25	26	27	28	29	30
13	14	15	16	17	18	19	Tammuz>	1	2	3	4	5	6	7
20	21	22	23	24	25	26		8	9	10	11	12	13	14
27	28	29	30	1	2	3		15	16	17	18	19	20	21
4	5	6	7	8	9	10	<July	22	23	24	25	26	27	28
11	12	13	14	15	16	17		29	1	2	3	4	5	6
18	19	20	21	22	23	24	Ab>	7	8	9	10	11	12	13
25	26	27	28	29	30	31		14	15	16	17	18	19	20
1	2	3	4	5	6	7	<August	21	22	23	24	25	26	27
8	9	10	11	12	13	14		28	29	30	1	2	3	4
15	16	17	18	19	20	21	Elul>	5	6	7	8	9	10	11
22	23	24	25	26	27	28		12	13	14	15	16	17	18
29	30	31	1	2	3	4	<Sept	19	20	21	22	23	24	25
5	6	7	8	9	10	11		26	27	28	29	1	2	3
12	13	14	15	16	17	18	Tishri>	4	5	6	7	8	9	10
19	20	21	22	23	24	25		11	12	13	14	15	16	17
26	27	28	29	30	1	2		18	19	20	21	22	23	24
3	4	5	6	7	8	9	<Oct	25	26	27	28	29	30	1
10	11	12	13	14	15	16	Heshvan>	2	3	4	5	6	7	8
17	18	19	20	21	22	23		9	10	11	12	13	14	15
24	25	26	27	28	29	30		16	17	18	19	20	21	22
31	1	2	3	4	5	6	<Nov	23	24	25	26	27	28	29
7	8	9	10	11	12	13	Kislev>	1	2	3	4	5	6	7
14	15	16	17	18	19	20		8	9	10	11	12	13	14
21	22	23	24	25	26	27		15	16	17	18	19	20	21
28	29	30	1	2	3	4		22	23	24	25	26	27	28
5	6	7	8	9	10	11	<Dec	29	30	1	2	3	4	5
12	13	14	15	16	17	18	Tebeth>	6	7	8	9	10	11	12
19	20	21	22	23	24	25		13	14	15	16	17	18	19
26	27	28	29	30	31			20	21	22	23	24	25	

AD 550/Dionysius' Cycle I, Year 19 (emb.)/Metonic Year 16/Easter 24 April, Luna 21

S	*M*	*T*	*W*	*T*	*F*	*S*		*S*	*M*	*T*	*W*	*T*	*F*	*S*
						1	Tebeth>							26
2	3	4	5	6	7	8	<Jan	27	28	29	1	2	3	4
9	10	11	12	13	14	15	Shevat>	5	6	7	8	9	10	11
16	17	18	19	20	21	22		12	13	14	15	16	17	18
23	24	25	26	27	28	29		19	20	21	22	23	24	25
30	31	1	2	3	4	5	<Feb	26	27	28	29	30	1	2
6	7	8	9	10	11	12	Adar>	3	4	5	6	7	8	9
13	14	15	16	17	18	19		10	11	12	13	14	15	16
20	21	22	23	24	25	26		17	18	19	20	21	22	23
27	28	1	2	3	4	5	<March	24	25	26	27	28	29	30
6	7	8	9	10	11	12	Adar 2>	1	2	3	4	5	6	7
13	14	15	16	17	18	19		8	9	10	11	12	13	14
20	21	22	23	24	25	26		15	16	17	18	19	20	21
27	28	29	30	31	1	2	<April	22	23	24	25	26	27	28
3	4	5	6	7	8	9	Nisan>	29	1	2	3	4	5	6
10	11	12	13	14	15	16		7	8	9	10	11	12	13
17	18	19	20	21	22	23		14	15	16	17	18	19	20
24	25	26	27	28	29	30		21	22	23	24	25	26	27
1	2	3	4	5	6	7	<May	28	29	30	1	2	3	4
8	9	10	11	12	13	14	Iyar>	5	6	7	8	9	10	11
15	16	17	18	19	20	21		12	13	14	15	16	17	18
22	23	24	25	26	27	28		19	20	21	22	23	24	25
29	30	31	1	2	3	4	Sivan>	26	27	28	29	1	2	3
5	6	7	8	9	10	11	<June	4	5	6	7	8	9	10
12	13	14	15	16	17	18		11	12	13	14	15	16	17
19	20	21	22	23	24	25		18	19	20	21	22	23	24
26	27	28	29	30	1	2	<July	25	26	27	28	29	30	1
3	4	5	6	7	8	9	Tammuz>	2	3	4	5	6	7	8
10	11	12	13	14	15	16		9	10	11	12	13	14	15
17	18	19	20	21	22	23		16	17	18	19	20	21	22
24	25	26	27	28	29	30		23	24	25	26	27	28	29
31	1	2	3	4	5	6	<August	1	2	3	4	5	6	7
7	8	9	10	11	12	13	Ab>	8	9	10	11	12	13	14
14	15	16	17	18	19	20		15	16	17	18	19	20	21
21	22	23	24	25	26	27		22	23	24	25	26	27	28
28	29	30	31	1	2	3	Elul>	29	30	1	2	3	4	5
4	5	6	7	8	9	10	<Sept	6	7	8	9	10	11	12
11	12	13	14	15	16	17		13	14	15	16	17	18	19
18	19	20	21	22	23	24		20	21	22	23	24	25	26
25	26	27	28	29	30	1	Tishri>	27	28	29	1	2	3	4
2	3	4	5	6	7	8	<Oct	5	6	7	8	9	10	11
9	10	11	12	13	14	15		12	13	14	15	16	17	18
16	17	18	19	20	21	22		19	20	21	22	23	24	25
23	24	25	26	27	28	29	Heshvan>	26	27	28	29	30	1	2
30	31	1	2	3	4	5	<Nov	3	4	5	6	7	8	9
6	7	8	9	10	11	12		10	11	12	13	14	15	16
13	14	15	16	17	18	19		17	18	19	20	21	22	23
20	21	22	23	24	25	26		24	25	26	27	28	29	1
27	28	29	30	1	2	3	*Kislev>	2	3	4	5	6	7	8
4	5	6	7	8	9	10	<Dec	9	10	11	12	13	14	15
11	12	13	14	15	16	17		16	17	18	19	20	21	22
18	19	20	21	22	23	24		23	24	25	26	27	28	1
25	26	27	28	29	30	31		2	3	4	5	6	7	8

AD 551/Dionysius' Cycle II, Year 1/Metonic Year 17/Easter 09 April, Luna 18

S	*M*	*T*	*W*	*T*	*F*	*S*		*S*	*M*	*T*	*W*	*T*	*F*	*S*
1	2	3	4	5	6	7	<Jan	9	10	11	12	13	14	15
8	9	10	11	12	13	14	Tebeth>	16	17	18	19	20	21	22
15	16	17	18	19	20	21		23	24	25	26	27	28	29
22	23	24	25	26	27	28	Shebat>	1	2	3	4	5	6	7
29	30	31	1	2	3	4	<Feb	8	9	10	11	12	13	14
5	6	7	8	9	10	11		15	16	17	18	19	20	21
12	13	14	15	16	17	18		22	23	24	25	26	27	28
19	20	21	22	23	24	25	Adar>	29	30	1	2	3	4	5
26	27	28	1	2	3	4	<March	6	7	8	9	10	11	12
5	6	7	8	9	10	11		13	14	15	16	17	18	19
12	13	14	15	16	17	18		20	21	22	23	24	25	26
19	20	21	22	23	24	25	Nisan>	27	28	29	30	1	2	3
26	27	28	29	30	31	1	<April	4	5	6	7	8	9	10
2	3	4	5	6	7	8		11	12	13	14	15	16	17
9	10	11	12	13	14	15		18	19	20	21	22	23	24
16	17	18	19	20	21	22		25	26	27	28	29	30	1
23	24	25	26	27	28	29	Iyar>	2	3	4	5	6	7	8
30	1	2	3	4	5	6	<May	9	10	11	12	13	14	15
7	8	9	10	11	12	13		16	17	18	19	20	21	22
14	15	16	17	18	19	20		23	24	25	26	27	28	29
21	22	23	24	25	26	27	Sivan>	1	2	3	4	5	6	7
28	29	30	31	1	2	3	<June	8	9	10	11	12	13	14
4	5	6	7	8	9	10		15	16	17	18	19	20	21
11	12	13	14	15	16	17		22	23	24	25	26	27	28
18	19	20	21	22	23	24	Tammuz>	29	30	1	2	3	4	5
25	26	27	28	29	30	1		6	7	8	9	10	11	12
2	3	4	5	6	7	8	<July	13	14	15	16	17	18	19
9	10	11	12	13	14	15		20	21	22	23	24	25	26
16	17	18	19	20	21	22	Ab>	27	28	29	1	2	3	4
23	24	25	26	27	28	29		5	6	7	8	9	10	11
30	31	1`	2	3	4	5	<August	12	13	14	15	16	17	18
6	7	8	9	10	11	12		19	20	21	22	23	24	25
13	14	15	16	17	18	19		26	27	28	29	30	1	2
20	21	22	23	24	25	26	Elul>	3	4	5	6	7	8	9
27	28	29	30	31	1	2	<Sept	10	11	12	13	14	15	16
3	4	5	6	7	8	9		17	18	19	20	21	22	23
10	11	12	13	14	15	16		24	25	26	27	28	29	1
17	18	19	20	21	22	23	Tishri>	2	3	4	5	6	7	8
24	25	26	27	28	29	30		9	10	11	12	13	14	15
1	2	3	4	5	6	7	<Oct	16	17	18	19	20	21	22
8	9	10	11	12	13	14		23	24	25	26	27	28	29
15	16	17	18	19	20	21	Heshvan>	30	1	2	3	4	5	6
22	23	24	25	26	27	28		7	8	9	10	11	12	13
29	30	31	1	2	3	4	<Nov	14	15	16	17	18	19	20
5	6	7	8	9	10	11		21	22	23	24	25	26	27
12	13	14	15	16	17	18	Kislev>	28	29	1	2	3	4	5
19	20	21	22	23	24	25		6	7	8	9	10	11	12
26	27	28	29	30	1	2	<Dec	13	14	15	16	17	18	19
3	4	5	6	7	8	9		20	21	22	23	24	25	26
10	11	12	13	14	15	16	Tebeth>	27	28	29	30	1	2	3
17	18	19	20	21	22	23		4	5	6	7	8	9	10
24	25	26	27	28	29	30		11	12	13	14	15	16	17
31								18						

AD 552/Dionysius' Cycle II, Year 2/Metonic Year 18/Easter 31 March, Luna 20

S	*M*	*T*	*W*	*T*	*F*	*S*		*S*	*M*	*T*	*W*	*T*	*F*	*S*
	1	2	3	4	5	6	Tebeth>		19	20	21	22	23	24
7	8	9	10	11	12	13	<Jan	25	26	27	28	29	1	2
14	15	16	17	18	19	20	Shevat>	3	4	5	6	7	8	9
21	22	23	24	25	26	27		10	11	12	13	14	15	16
28	29	30	31	1	2	3	<Feb	17	18	19	20	21	22	23
4	5	6	7	8	9	10		24	25	26	27	28	29	30
11	12	13	14	15	16	17	Adar>	1	2	3	4	5	6	7
18	19	20	21	22	23	24		8	9	10	11	12	13	14
25	26	27	28	29	1	2	<Mar	15	16	17	18	19	20	21
3	4	5	6	7	8	9		22	23	24	25	26	27	28
10	11	12	13	14	15	16	Nisan>	29	30	1	2	3	4	5
17	18	19	20	21	22	23		6	7	8	9	10	11	12
24	25	26	27	28	29	30		13	14	15	16	17	18	19
31	1	2	3	4	5	6	<April	20	21	22	23	24	25	26
7	8	9	10	11	12	13	Iyar>	27	28	29	30	1	2	3
14	15	16	17	18	19	20		4	5	6	7	8	9	10
21	22	23	24	25	26	27		11	12	13	14	15	16	17
28	29	30	1	2	3	4	<May	18	19	20	21	22	23	24
5	6	7	8	9	10	11	Sivan>	25	26	27	28	29	1	2
12	13	14	15	16	17	18		3	4	5	6	7	8	9
19	20	21	22	23	24	25		10	11	12	13	14	15	16
26	27	28	29	30	31	1	<June	17	18	19	20	21	22	23
2	3	4	5	6	7	8		24	25	26	27	28	29	30
9	10	11	12	13	14	15	Tammuz>	1	2	3	4	5	6	7
16	17	18	19	20	21	22		8	9	10	11	12	13	14
23	24	25	26	27	28	29		15	16	17	18	19	20	21
30	1	2	3	4	5	6	<July	22	23	24	25	26	27	28
7	8	9	10	11	12	13	Ab>	29	1	2	3	4	5	6
14	15	16	17	18	19	20		7	8	9	10	11	12	13
21	22	23	24	25	26	27		14	15	16	17	18	19	20
28	29	30	31	1	2	3	<August	21	22	23	24	25	26	27
4	5	6	7	8	9	10	Elul>	28	29	30	1	2	3	4
11	12	13	14	15	16	17		5	6	7	8	9	10	11
18	19	20	21	22	23	24		12	13	14	15	16	17	18
25	26	27	28	29	30	31		19	20	21	22	23	24	25
1	2	3	4	5	6	7	<Sept	26	27	28	29	1	2	3
8	9	10	11	12	13	14	Tishri>	4	5	6	7	8	9	10
15	16	17	18	19	20	21		11	12	13	14	15	16	17
22	23	24	25	26	27	28		18	19	20	21	22	23	24
29	30	1	2	3	4	5	<Oct	25	26	27	28	29	30	1
6	7	8	9	10	11	12	Heshvan>	2	3	4	5	6	7	8
13	14	15	16	17	18	19		9	10	11	12	13	14	15
20	21	22	23	24	25	26		16	17	18	19	20	21	22
27	28	29	30	31	1	2	<Nov	23	24	25	26	27	28	29
3	4	5	6	7	8	9	Kislev>	1	2	3	4	5	6	7
10	11	12	13	14	15	16		8	9	10	11	12	13	14
17	18	19	20	21	22	23		15	16	17	18	19	20	21
24	25	26	27	28	29	30		22	23	24	25	26	27	28
1	2	3	4	5	6	7	<Dec	29	30	1	2	3	4	5
8	9	10	11	12	13	14	Tebeth>	6	7	8	9	10	11	12
15	16	17	18	19	20	21		13	14	15	16	17	18	19
22	23	24	25	26	27	28		20	21	22	23	24	25	26
29	30	31						27	28	29				

AD 553/Dionysius' Cycle II, Year 3 (emb.)/Metonic Year 19/Easter 20 April, Luna 21

S	*M*	*T*	*W*	*T*	*F*	*S*		*S*	*M*	*T*	*W*	*T*	*F*	*S*
			1	2	3	4	<Jan				1	2	3	4
5	6	7	8	9	10	11	Shebat>	5	6	7	8	9	10	11
12	13	14	15	16	17	18		12	13	14	15	16	17	18
19	20	21	22	23	24	25		19	20	21	22	23	24	25
26	27	28	29	30	31	1	Adar>	26	27	28	29	30	1	2
2	3	4	5	6	7	8	<Feb	3	4	5	6	7	8	9
9	10	11	12	13	14	15		10	11	12	13	14	15	16
16	17	18	19	20	21	22		17	18	19	20	21	22	23
23	24	25	26	27	28	1		24	25	26	27	28	29	30
2	3	4	5	6	7	8	<March	1	2	3	4	5	6	7
9	10	11	12	13	14	15	Adar 2>	8	9	10	11	12	13	14
16	17	18	19	20	21	22		15	16	17	18	19	20	21
23	24	25	26	27	28	29		22	23	24	25	26	27	28
30	31	1	2	3	4	5	<April	29	1	2	3	4	5	6
6	7	8	9	10	11	12	Nisan>	7	8	9	10	11	12	13
13	14	15	16	17	18	19		14	15	16	17	18	19	20
20	21	22	23	24	25	26		21	22	23	24	25	26	27
27	28	29	30	1	2	3	<May	28	29	30	1	2	3	4
4	5	6	7	8	9	10	Iyar>	5	6	7	8	9	10	11
11	12	13	14	15	16	17		12	13	14	15	16	17	18
18	19	20	21	22	23	24		19	20	21	22	23	24	25
25	26	27	28	29	30	31	Sivan>	26	27	28	29	1	2	3
1	2	3	4	5	6	7	<June	4	5	6	7	8	9	10
8	9	10	11	12	13	14		11	12	13	14	15	16	17
15	16	17	18	19	20	21		18	19	20	21	22	23	24
22	23	24	25	26	27	28	Tammuz>	25	26	27	28	29	30	1
29	30	1	2	3	4	5	<July	2	3	4	5	6	7	8
6	7	8	9	10	11	12		9	10	11	12	13	14	15
13	14	15	16	17	18	19		16	17	18	19	20	21	22
20	21	22	23	24	25	26		23	24	25	26	27	28	29
27	28	29	30	31	1	2	Ab>	1	2	3	4	5	6	7
3	4	5	6	7	8	9	<August	8	9	10	11	12	13	14
10	11	12	13	14	15	16		15	16	17	18	19	20	21
17	18	19	20	21	22	23		22	23	24	25	26	27	28
24	25	26	27	28	29	30	Elul>	29	30	1	2	3	4	5
31	1	2	3	4	5	6	<Sept	6	7	8	9	10	11	12
7	8	9	10	11	12	13		13	14	15	16	17	18	19
14	15	16	17	18	19	20		20	21	22	23	24	25	26
21	22	23	24	25	26	27	Tishri>	27	28	29	1	2	3	4
28	29	30	1	2	3	4	<Oct	5	6	7	8	9	10	11
5	6	7	8	9	10	11		12	13	14	15	16	17	18
12	13	14	15	16	17	18		19	20	21	22	23	24	25
19	20	21	22	23	24	25		26	27	28	29	30	1	2
26	27	28	29	30	31	1	Heshvan>	3	4	5	6	7	8	9
2	3	4	5	6	7	8	<Nov	10	11	12	13	14	15	16
9	10	11	12	13	14	15		17	18	19	20	21	22	23
16	17	18	19	20	21	22		24	25	26	27	28	29	1
23	24	25	26	27	28	29	Kislev>	2	3	4	5	6	7	8
30	1	2	3	4	5	6	<Dec	9	10	11	12	13	14	15
7	8	9	10	11	12	13		16	17	18	19	20	21	22
14	15	16	17	18	19	20		23	24	25	26	27	28	29
21	22	23	24	25	26	27	Tebeth>	1	2	3	4	5	6	7
28	29	30	31					8	9	10	11			

AD 554/Dionysius; Cycle II, Year 4/Metonic Year 1/Easter 05 April, Luna 17

S	*M*	*T*	*W*	*T*	*F*	*S*		*S*	*M*	*T*	*W*	*T*	*F*	*S*
				1	2	3	Tebeth>					12	13	14
4	5	6	7	8	9	10	<Jan	15	16	17	18	19	20	21
11	12	13	14	15	16	17		22	23	24	25	26	27	28
18	19	20	21	22	23	24	Shevat>	29	1	2	3	4	5	6
25	26	27	28	29	30	31		7	8	9	10	11	12	13
1	2	3	4	5	6	7	<Feb	14	15	16	17	18	19	20
8	9	10	11	12	13	14		21	22	23	24	25	26	27
15	16	17	18	19	20	21	Adar>	28	29	30	1	2	3	4
22	23	24	25	26	27	28		5	6	7	8	9	10	11
1	2	3	4	5	6	7	<March	12	13	14	15	16	17	18
8	9	10	11	12	13	14		19	20	21	22	23	24	25
15	16	17	18	19	20	21		26	27	28	29	30	1	2
22	23	24	25	26	27	28	Nisan>	3	4	5	6	7	8	9
29	30	31	1	2	3	4		10	11	12	13	14	15	16
5	6	7	8	9	10	11	<April	17	18	19	20	21	22	23
12	13	14	15	16	17	18		24	25	26	27	28	29	30
19	20	21	22	23	24	25	Iyar>	1	2	3	4	5	6	7
26	27	28	29	30	1	2	<May	8	9	10	11	12	13	14
3	4	5	6	7	8	9		15	16	17	18	19	20	21
10	11	12	13	14	15	16		22	23	24	25	26	27	28
17	18	19	20	21	22	23	Sivan>	29	1	2	3	4	5	6
24	25	26	27	28	29	30		7	8	9	10	11	12	13
31	1	2	3	4	5	6	<June	14	15	16	17	18	19	20
7	8	9	10	11	12	13		21	22	23	24	25	26	27
14	15	16	17	18	19	20	Tammuz>	28	29	30	1	2	3	4
21	22	23	24	25	26	27		5	6	7	8	9	10	11
28	29	30	1	2	3	4	<July	12	13	14	15	16	17	18
5	6	7	8	9	10	11		19	20	21	22	23	24	25
12	13	14	15	16	17	18	Ab>	26	27	28	29	1	2	3
19	20	21	22	23	24	25		4	5	6	7	8	9	10
26	27	28	29	30	31	1		11	12	13	14	15	16	17
2	3	4	5	6	7	8	<August	18	19	20	21	22	23	24
9	10	11	12	13	14	15		25	26	27	28	29	30	1
16	17	18	19	20	21	22	Elul>	2	3	4	5	6	7	8
23	24	25	26	27	28	29		9	10	11	12	13	14	15
30	31	1	2	3	4	5	<Sept	16	17	18	19	20	21	22
6	7	8	9	10	11	12		23	24	25	26	27	28	29
13	14	15	16	17	18	19	Tishri>	1	2	3	4	5	6	7
20	21	22	23	24	25	26		8	9	10	11	12	13	14
27	28	29	30	1	2	3	<Oct	15	16	17	18	19	20	21
4	5	6	7	8	9	10		22	23	24	25	26	27	28
11	12	13	14	15	16	17	Heshvan>	29	30	1	2	3	4	5
18	19	20	21	22	23	24		6	7	8	9	10	11	12
25	26	27	28	29	30	31		13	14	15	16	17	18	19
1	2	3	4	5	6	7	<Nov	20	21	22	23	24	25	26
8	9	10	11	12	13	14		27	28	29	1	2	3	4
15	16	17	18	19	20	21		5	6	7	8	9	10	11
22	23	24	25	26	27	28		12	13	14	15	16	17	18
29	30	1	2	3	4	5	<Dec	19	20	21	22	23	24	25
6	7	8	9	10	11	12	Tebeth>	26	27	28	29	1	2	3
13	14	15	16	17	18	19		4	5	6	7	8	9	10
20	21	22	23	24	25	26		11	12	13	14	15	16	17
27	28	29	30	31				18	19	20	21	22		

AD 555/Dionysius' Cycle II, Year 5/Metonic Year 2/Easter 28 March, Luna 20

S	*M*	*T*	*W*	*T*	*F*	*S*		*S*	*M*	*T*	*W*	*T*	*F*	*S*
					1	2	Tebeth>						23	24
3	4	5	6	7	8	9	<Jan	25	26	27	28	29	1	2
10	11	12	13	14	15	16	Shevat>	3	4	5	6	7	8	9
17	18	19	20	21	22	23		10	11	12	13	14	15	16
24	25	26	27	28	29	30		17	18	19	20	21	22	23
31	1	2	3	4	5	6	<Feb	24	25	26	27	28	29	30
7	8	9	10	11	12	13	Adar>	1	2	3	4	5	6	7
14	15	16	17	18	19	20		8	9	10	11	12	13	14
21	22	23	24	25	26	27		15	16	17	18	19	20	21
28	1	2	3	4	5	6	<March	22	23	24	25	26	27	28
7	8	9	10	11	12	13	Nisan>	29	30	1	2	3	4	5
14	15	16	17	18	19	20		6	7	8	9	10	11	12
21	22	23	24	25	26	27		13	14	15	16	17	18	19
28	29	30	31	1	2	3	<April	20	21	22	23	24	25	26
4	5	6	7	8	9	10		27	28	29	30	1	2	3
11	12	13	14	15	16	17	Iyar>	4	5	6	7	8	9	10
18	19	20	21	22	23	24		11	12	13	14	15	16	17
25	26	27	28	29	30	1		18	19	20	21	22	23	24
2	3	4	5	6	7	8	<May	25	26	27	28	29	1	2
9	10	11	12	13	14	15	Sivan>	3	4	5	6	7	8	9
16	17	18	19	20	21	22		10	11	12	13	14	15	16
23	24	25	26	27	28	29		17	18	19	20	21	22	23
30	31	1	2	3	4	5	<June	24	25	26	27	28	29	30
6	7	8	9	10	11	12	Tammuz>	1	2	3	4	5	6	7
13	14	15	16	17	18	19		8	9	10	11	12	13	14
20	21	22	23	24	25	26		15	16	17	18	19	20	21
27	28	29	30	1	2	3	<July	22	23	24	25	26	27	28
4	5	6	7	8	9	10	Ab>	29	1	2	3	4	5	6
11	12	13	14	15	16	17		7	8	9	10	11	12	13
18	19	20	21	22	23	24		14	15	16	17	18	19	20
25	26	27	28	29	30	31		21	22	23	24	25	26	27
1	2	3	4	5	6	7	<August	28	29	30	1	2	3	4
8	9	10	11	12	13	14	Elul>	5	6	7	8	9	10	11
15	16	17	18	19	20	21		12	13	14	15	16	17	18
22	23	24	25	26	27	28		19	20	21	22	23	24	25
29	30	31	1	2	3	4	<Sept	26	27	28	29	1	2	3
5	6	7	8	9	10	11	Tishri>	4	5	6	7	8	9	10
12	13	14	15	16	17	18		11	12	13	14	15	16	17
19	20	21	22	23	24	25		18	19	20	21	22	23	24
26	27	28	29	30	1	2	<Oct	25	26	27	28	29	30	1
3	4	5	6	7	8	9	Heshvan>	2	3	4	5	6	7	8
10	11	12	13	14	15	16		9	10	11	12	13	14	15
17	18	19	20	21	22	23		16	17	18	19	20	21	22
24	25	26	27	28	29	30		23	24	25	26	27	28	29
31	1	2	3	4	5	6	<Nov	30	1	2	3	4	5	6
7	8	9	10	11	12	13	Kislev>	7	8	9	10	11	12	13
14	15	16	17	18	19	20		14	15	16	17	18	19	20
21	22	23	24	25	26	27		21	22	23	24	25	26	27
28	29	30	1	2	3	4	<Dec	28	29	30	1	2	3	4
5	6	7	8	9	10	11	Tebeth>	5	6	7	8	9	10	11
12	13	14	15	16	17	18		12	13	14	15	16	17	18
19	20	21	22	23	24	25		19	20	21	22	23	24	25
26	27	28	29	30	31		Shevat>	26	27	28	29	1	2	

AD 556/Dionysius' Cycle II, Year 6 (emb.)/Metonic Year 3/Easter 16 April, Luna 20

S	*M*	*T*	*W*	*T*	*F*	*S*		*S*	*M*	*T*	*W*	*T*	*F*	*S*
2	3	4	5	6	7	1/8	<Jan	4	5	6	7	8	9	3/10
9	10	11	12	13	14	15	Shevat>	11	12	13	14	15	16	17
16	17	18	19	20	21	22		18	19	20	21	22	23	24
23	24	25	26	27	28	29	Adar>	25	26	27	28	29	30	1
30	31	1	2	3	4	5	<Feb	2	3	4	5	6	7	8
6	7	8	9	10	11	12		9	10	11	12	13	14	15
13	14	15	16	17	18	19		16	17	18	19	20	21	22
20	21	22	23	24	25	26		23	24	25	26	27	28	29
27	28	29	1	2	3	4	<March	30	1	2	3	4	5	6
5	6	7	8	9	10	11	Adar 2>	7	8	9	10	11	12	13
12	13	14	15	16	17	18		14	15	16	17	18	19	20
19	20	21	22	23	24	25		21	22	23	24	25	26	27
26	27	28	29	30	31	1	Nisan>	28	29	1	2	3	4	5
2	3	4	5	6	7	8	<April	6	7	8	9	10	11	12
9	10	11	12	13	14	15		13	14	15	16	17	18	19
16	17	18	19	20	21	22		20	21	22	23	24	25	26
23	24	25	26	27	28	29	Iyar>	27	28	29	30	1	2	3
30	1	2	3	4	5	6	<May	4	5	6	7	8	9	10
7	8	9	10	11	12	13		11	12	13	14	15	16	17
14	15	16	17	18	19	20		18	19	20	21	22	23	24
21	22	23	24	25	26	27	Sivan>	25	26	27	28	29	1	2
28	29	30	31	1	2	3	<June	3	4	5	6	7	8	9
4	5	6	7	8	9	10		10	11	12	13	14	15	16
11	12	13	14	15	16	17		17	18	19	20	21	22	23
18	19	20	21	22	23	24		24	25	26	27	28	29	30
25	26	27	28	29	30	1	Tammuz>	1	2	3	4	5	6	7
2	3	4	5	6	7	8	<July	8	9	10	11	12	13	14
9	10	11	12	13	14	15		15	16	17	18	19	20	21
16	17	18	19	20	21	22		22	23	24	25	26	27	28
23	24	25	26	27	28	29	Ab>	29	1	2	3	4	5	6
30	31	1	2	3	4	5	<August	7	8	9	10	11	12	13
6	7	8	9	10	11	12		14	15	16	17	18	19	20
13	14	15	16	17	18	19		21	22	23	24	25	26	27
20	21	22	23	24	25	26		28	29	30	1	2	3	4
27	28	29	30	31	1	2	Elul>	5	6	7	8	9	10	11
3	4	5	6	7	8	9	<Sept	12	13	14	15	16	17	18
10	11	12	13	14	15	16		19	20	21	22	23	24	25
17	18	19	20	21	22	23		26	27	28	29	1	2	3
24	25	26	27	28	29	30	Tishri>	4	5	6	7	8	9	10
1	2	3	4	5	6	7	<Oct	11	12	13	14	15	16	17
8	9	10	11	12	13	14		18	19	20	21	22	23	24
15	16	17	18	19	20	21		25	26	27	28	29	30	1
22	23	24	25	26	27	28	Heshvan>	2	3	4	5	6	7	8
29	30	31	1	2	3	4	<Nov	9	10	11	12	13	14	15
5	6	7	8	9	10	11		16	17	18	19	20	21	22
12	13	14	15	16	17	18		23	24	25	26	27	28	29
19	20	21	22	23	24	25	Kislev>	1	2	3	4	5	6	7
26	27	28	29	30	1	2	<Dec	8	9	10	11	12	13	14
3	4	5	6	7	8	9		15	16	17	18	19	20	21
10	11	12	13	14	15	16		22	23	24	25	26	27	28
17	18	19	20	21	22	23	Tebeth>	29	1	2	3	4	5	6
24	25	26	27	28	29	30		7	8	9	10	11	12	13
31								14						

AD 557/Dionysius' Cycle II, Year 7/Metonic Year 4/Easter 01 April, Luna 16

S	*M*	*T*	*W*	*T*	*F*	*S*		*S*	*M*	*T*	*W*	*T*	*F*	*S*
	1	2	3	4	5	6	Tebeth>		15	16	17	18	19	20
7	8	9	10	11	12	13	<Jan	21	22	23	24	25	26	27
14	15	16	17	18	19	20	Shevat>	28	29	1	2	3	4	5
21	22	23	24	25	26	27		6	7	8	9	10	11	12
28	29	30	31	1	2	3	<Feb	13	14	15	16	17	18	19
4	5	6	7	8	9	10		20	21	22	23	24	25	26
11	12	13	14	15	16	17	Adar>	27	18	29	30	1	2	3
18	19	20	21	22	23	24		4	5	6	7	8	9	10
25	26	27	28	1	2	3	<March	11	12	13	14	15	16	17
4	5	6	7	8	9	10		18	19	20	21	22	23	24
11	12	13	14	15	16	17		25	26	27	28	29	30	1
18	19	20	21	22	23	24	Nisan>	2	3	4	5	6	7	8
25	26	27	28	29	30	31		9	10	11	12	13	14	15
1	2	3	4	5	6	7	<April	16	17	18	19	20	21	22
8	9	10	11	12	13	14		23	24	25	26	27	28	29
15	16	17	18	19	20	21	Iyar>	30	1	2	3	4	5	6
22	23	24	25	26	27	28		7	8	9	10	11	12	13
29	30	1	2	3	4	5	<May	14	15	16	17	18	19	20
6	7	8	9	10	11	12		21	22	23	24	25	26	27
13	14	15	16	17	18	19	Sivan>	28	29	1	2	3	4	5
20	21	22	23	24	25	26		6	7	8	9	10	11	12
27	28	29	30	31	1	2	<June	13	14	15	16	17	18	19
3	4	5	6	7	8	9		20	21	22	23	24	25	26
10	11	12	13	14	15	16	Tammuz>	27	28	29	30	1	2	3
17	18	19	20	21	22	23		4	5	6	7	8	9	10
24	25	26	27	28	29	30		11	12	13	14	15	16	17
1	2	3	4	5	6	7	<July	18	19	20	21	22	23	24
8	9	10	11	12	13	14		25	26	27	28	29	1	2
15	16	17	18	19	20	21	Ab>	3	4	5	6	7	8	9
22	23	24	25	26	27	28		10	11	12	13	14	15	16
29	30	31	1	2	3	4	<August	17	18	19	20	21	22	23
5	6	7	8	9	10	11		24	25	26	27	28	29	30
12	13	14	15	16	17	18	Elul>	1	2	3	4	5	6	7
19	20	21	22	23	24	25		8	9	10	11	12	13	14
26	27	28	29	30	31	1		15	16	17	18	19	20	21
2	3	4	5	6	7	8	<Sept	22	23	24	25	26	27	28
9	10	11	12	13	14	15	Tishri>	29	1	2	3	4	5	6
16	17	18	19	20	21	22		7	8	9	10	11	12	13
23	24	25	26	27	28	29		14	15	16	17	18	19	20
30	1	2	3	4	5	6	<Oct	21	22	23	24	25	26	27
7	8	9	10	11	12	13	Heshvan>	28	29	30	1	2	3	4
14	15	16	17	18	19	20		5	6	7	8	9	10	11
21	22	23	24	25	26	27		12	13	14	15	16	17	18
28	29	30	31	1	2	3	<Nov	19	20	21	22	23	24	25
4	5	6	7	8	9	10	Kislev>	26	27	28	29	1	2	3
11	12	13	14	15	16	17		4	5	6	7	8	9	10
18	19	20	21	22	23	24		11	12	13	14	15	16	17
25	26	27	28	29	30	1		18	19	20	21	22	23	24
2	3	4	5	6	7	8	<Dec	25	26	27	28	29	30	1
9	10	11	12	13	14	15	Tebeth>	2	3	4	5	6	7	8
16	17	18	19	20	21	22		9	10	11	12	13	14	15
23	24	25	26	27	28	29		16	17	18	19	20	21	22
30	31							23	24					

AD 558/Dionysius' Cycle II, Year 8 (emb.)/Metonic Year 5/Easter 21 April, Luna 17

S	*M*	*T*	*W*	*T*	*F*	*S*		*S*	*M*	*T*	*W*	*T*	*F*	*S*
		1	2	3	4	5	Tebeth>			25	26	27	28	29
6	7	8	9	10	11	12	<Jan	1	2	3	4	5	6	7
13	14	15	16	17	18	19		8	9	10	11	12	13	14
20	21	22	23	24	25	26	6	15	16	17	18	19	20	21
27	28	29	30	31	1	2	<Feb	22	23	24	25	26	27	28
3	4	5	6	7	8	9	Shevat>	29	30	1	2	3	4	5
10	11	12	13	14	15	16		6	7	8	9	10	11	12
17	18	19	20	21	22	23		13	14	15	16	17	18	19
24	25	26	27	28	1	2	<March	20	21	22	23	24	25	26
3	4	5	6	7	8	9	Adar 2>	27	28	29	30	1	2	3
10	11	12	13	14	15	16		4	5	6	7	8	9	10
17	18	19	20	21	22	23		11	12	13	14	15	16	17
24	25	26	27	28	29	30		18	19	20	21	22	23	24
31	1	2	3	4	5	6	<April	25	26	27	28	29	1	2
7	8	9	10	11	12	13	Nisan>	3	4	5	6	7	8	9
14	15	16	17	18	19	20		10	11	12	13	14	15	16
21	22	23	24	25	26	27		17	18	19	20	21	22	23
28	29	30	1	2	3	4	<May	24	25	26	27	28	29	30
5	6	7	8	9	10	11	Iyar>	1	2	3	4	5	6	7
12	13	14	15	16	17	18		8	9	10	11	12	13	14
19	20	21	22	23	24	25		15	16	17	18	19	20	21
26	27	28	29	30	31	1	<June	22	23	24	25	26	27	28
2	3	4	5	6	7	8	Sivan>	29	1	2	3	4	5	6
9	10	11	12	13	14	15		7	8	9	10	11	12	13
16	17	18	19	20	21	22		14	15	16	17	18	19	20
23	24	25	26	27	28	29		21	22	23	24	25	26	27
30	1	2	3	4	5	6	Tammuz>	28	29	30	1	2	3	4
7	8	9	10	11	12	13	<July	5	6	7	8	9	10	11
14	15	16	17	18	19	20		12	13	14	15	16	17	18
21	22	23	24	25	26	27		19	20	21	22	23	24	25
28	29	30	31	1	2	3	<August	26	27	28	29	1	2	3
4	5	6	7	8	9	10	Ab>	4	5	6	7	8	9	10
11	12	13	14	15	16	17		11	12	13	14	15	16	17
18	19	20	21	22	23	24		18	19	20	21	22	23	24
25	26	27	28	29	30	31	Elul>	25	26	27	28	29	30	1
1	2	3	4	5	6	7	<Sept	2	3	4	5	6	7	8
8	9	10	11	12	13	14		9	10	11	12	13	14	15
15	16	17	18	19	20	21		16	17	18	19	20	21	22
22	23	24	25	26	27	28		23	24	25	26	27	28	29
29	30	1	2	3	4	5	Tishri>	1	2	3	4	5	6	7
6	7	8	9	10	11	12	<Oct	8	9	10	11	12	13	14
13	14	15	16	17	18	19		15	16	17	18	19	20	21
20	21	22	23	24	25	26		22	23	24	25	26	27	28
27	28	29	30	31	1	2	<Nov	29	30	1	2	3	4	5
3	4	5	6	7	8	9	Hesvan>	6	7	8	9	10	11	12
10	11	12	13	14	15	16		13	14	15	16	17	18	19
17	18	19	20	21	22	23		20	21	22	23	24	25	26
24	25	26	27	28	29	30	Kislev>	27	28	29	1	2	3	4
1	2	3	4	5	6	7	<Dec	5	6	7	8	9	10	11
8	9	10	11	12	13	14		12	13	14	15	16	17	18
15	16	17	18	19	20	21		19	20	21	22	23	24	25
22	23	24	25	26	27	28	Tebeth>	26	27	28	29	1	2	3
29	30	31						4	5	6				

AD 559/Dionysius' Cycle II, Year 9/Metonic Year 6/Easter 13 April, Luna 20

S	*M*	*T*	*W*	*T*	*F*	*S*		*S*	*M*	*T*	*W*	*T*	*F*	*S*
			1	2	3	4	Tebeth>				7	8	9	10
5	6	7	8	9	10	11	<Jan	11	12	13	14	15	16	17
12	13	14	15	16	17	18		18	19	20	21	22	23	24
19	20	21	22	23	24	25		25	26	27	28	29	1	2
26	27	28	29	30	31	1	Shevat>	3	4	5	6	7	8	9
2	3	4	5	6	7	8	<Feb	10	11	12	13	14	15	16
9	10	11	12	13	14	15		17	18	19	20	21	22	23
16	17	18	19	20	21	22		24	25	26	27	28	29	30
23	24	25	26	27	28	1	Adar>	1	2	3	4	5	6	7
2	3	4	5	6	7	8	<March	8	9	10	11	12	13	14
9	10	11	12	13	14	15		15	16	17	18	19	20	21
16	17	18	19	20	21	22		22	23	24	25	26	27	28
23	24	25	26	27	28	29	Nisan>	29	30	1	2	3	4	5
30	31	1	2	3	4	5	<April	6	7	8	9	10	11	12
6	7	8	9	10	11	12		13	14	15	16	17	18	19
13	14	15	16	17	18	19		20	21	22	23	24	25	26
20	21	22	23	24	25	26	Iyar>	27	28	29	30	1	2	3
27	28	29	30	1	2	3	<May	4	5	6	7	8	9	10
4	5	6	7	8	9	10		11	12	13	14	15	16	17
11	12	13	14	15	16	17		18	19	20	21	22	23	24
18	19	20	21	22	23	24		25	26	27	28	29	1	2
25	26	27	28	29	30	31	Sivan>	3	4	5	6	7	8	9
1	2	3	4	5	6	7	<June	10	11	12	13	14	15	16
8	9	10	11	12	13	14		17	18	19	20	21	22	23
15	16	17	18	19	20	21		24	25	26	27	28	29	30
22	23	24	25	26	27	28	Tammuz>	1	2	3	4	5	6	7
29	30	1	2	3	4	5	<July	8	9	10	11	12	13	14
6	7	8	9	10	11	12		15	16	17	18	19	20	21
13	14	15	16	17	18	19		22	23	24	25	26	27	28
20	21	22	23	24	25	26	Ab>	29	1	2	3	4	5	6
27	28	29	30	31	1	2	<August	7	8	9	10	11	12	13
3	4	5	6	7	8	9		14	15	16	17	18	19	20
10	11	12	13	14	15	16		21	22	23	24	25	26	27
17	18	19	20	21	22	23	Elul>	28	29	30	1	2	3	4
24	25	26	27	28	29	30		5	6	7	8	9	10	11
31	1	2	3	4	5	6	<Sept	12	13	14	15	16	17	18
7	8	9	10	11	12	13		19	20	21	22	23	24	25
14	15	16	17	18	19	20	Tishri>	26	27	28	29	1	2	3
21	22	23	24	25	26	27		4	5	6	7	8	9	10
28	29	30	1	2	3	4	<Oct	11	12	13	14	15	16	17
5	6	7	8	9	10	11		18	19	20	21	22	23	24
12	13	14	15	16	17	18		25	26	27	28	29	30	1
19	20	21	22	23	24	25	Heshvan>	2	3	4	5	6	7	8
26	27	28	29	30	31	1		9	10	11	12	13	14	15
2	3	4	5	6	7	8	<Nov	16	17	18	19	20	21	22
9	10	11	12	13	14	15		23	24	25	26	27	28	29
16	17	18	19	20	21	22	Kislev>	1	2	3	4	5	6	7
23	24	25	26	27	28	29		8	9	10	11	12	13	14
30	1	2	3	4	5	6	<Dec	15	16	17	18	19	20	21
7	8	9	10	11	12	13		22	23	24	25	26	27	28
14	15	16	17	18	19	20	Tebeth>	29	30	1	2	3	4	5
21	22	23	24	25	26	27		6	7	8	9	10	11	12
28	29	30	31					13	14	15	16			

AD 560/Dionysius' Cycle II, Year 10/Metonic Year 7/Easter 28 March, Luna 15

S	M	T	W	T	F	S		S	M	T	W	T	F	S
				1	2	3	Tebeth>					17	18	19
4	5	6	7	8	9	10	<Jan	20	21	22	23	24	25	26
11	12	13	14	15	16	17	Shevat>	27	28	29	1	2	3	4
18	19	20	21	22	23	24		5	6	7	8	9	10	11
25	26	27	28	29	30	31		12	13	14	15	16	17	18
1	2	3	4	5	6	7	<Feb	19	20	21	22	23	24	25
8	9	10	11	12	13	14		26	27	28	29	30	1	2
15	16	17	18	19	20	21	Adar>	3	4	5	6	7	8	9
22	23	24	25	26	27	28		10	11	12	13	14	15	16
29	1	2	3	4	5	6	<March	17	18	19	20	21	22	23
7	8	9	10	11	12	13		24	25	26	27	28	29	30
14	15	16	17	18	19	20	Nisan>	1	2	3	4	5	6	7
21	22	23	24	25	26	27		8	9	10	11	12	13	14
28	29	30	31	1	2	3	<April	15	16	17	18	19	20	21
4	5	6	7	8	9	10		22	23	24	25	26	27	28
11	12	13	14	15	16	17	Iyar>	29	30	1	2	3	4	5
18	19	20	21	22	23	24		6	7	8	9	10	11	12
25	26	27	28	29	30	1		13	14	15	16	17	18	19
2	3	4	5	6	7	8	<May	20	21	22	23	24	25	26
9	10	11	12	13	14	15		27	28	29	1	2	3	4
16	17	18	19	20	21	22	Sivan>	5	6	7	8	9	10	11
23	24	25	26	27	28	29		12	13	14	15	16	17	18
30	31	1	2	3	4	5	<June	19	20	21	22	23	24	25
6	7	8	9	10	11	12		26	27	28	29	30	1	2
13	14	15	16	17	18	19	Tammuz>	3	4	5	6	7	8	9
20	21	22	23	24	25	26		10	11	12	13	14	15	16
27	28	29	30	1	2	3		17	18	19	20	21	22	23
4	5	6	7	8	9	10	<July	24	25	26	27	28	29	1
11	12	13	14	15	16	17	Ab>	2	3	4	5	6	7	8
18	19	20	21	22	23	24		9	10	11	12	13	14	15
25	26	27	28	29	30	31		16	17	18	19	20	21	22
1	2	3	4	5	6	7	<August	23	24	25	26	27	28	29
8	9	10	11	12	13	14	Elul>	30	1	2	3	4	5	6
15	16	17	18	19	20	21		7	8	9	10	11	12	13
22	23	24	25	26	27	28		14	15	16	17	18	19	20
29	30	31	1	2	3	4	<Sept	21	22	23	24	25	26	27
5	6	7	8	9	10	11	Tishri>	28	29	1	2	3	4	5
12	13	14	15	16	17	18		6	7	8	9	10	11	12
19	20	21	22	23	24	25		13	14	15	16	17	18	19
26	27	28	29	30	1	2	<Oct	20	21	22	23	24	25	26
3	4	5	6	7	8	9	Heshvan>	27	28	29	30	1	2	3
10	11	12	13	14	15	16		4	5	6	7	8	9	10
17	18	19	20	21	22	23		11	12	13	14	15	16	17
24	25	26	27	28	29	30		18	19	20	21	22	23	24
31	1	2	3	4	5	6	<Nov	25	26	27	28	29	1	2
7	8	9	10	11	12	13	Kislev>	3	4	5	6	7	8	9
14	15	16	17	18	19	20		10	11	12	13	14	15	16
21	22	23	24	25	26	27		17	18	19	20	21	22	23
28	29	30	1	2	3	4	<Dec	24	25	26	27	28	29	30
5	6	7	8	9	10	11	Tebeth>	1	2	3	4	5	6	7
12	13	14	15	16	17	18		8	9	10	11	12	13	14
19	20	21	22	23	24	25		15	16	17	18	19	20	21
26	27	28	29	30	31			22	23	24	25	26	27	

AD 561/Dionysius' Cycle II, Year 11 (emb.)/Metonic Year 8/Easter 17 April, Luna 16

S	*M*	*T*	*W*	*T*	*F*	*S*		*S*	*M*	*T*	*W*	*T*	*F*	*S*
						1	Tebeth>							28
2	3	4	5	6	7	8	<Jan	29	1	2	3	4	5	6
9	10	11	12	13	14	15	Shevat>	7	8	9	10	11	12	13
16	17	18	19	20	21	22		14	15	16	17	18	19	20
23	24	25	26	27	28	29		21	22	23	24	25	26	27
30	31	1	2	3	4	5	<Feb	28	29	30	1	2	3	4
6	7	8	9	10	11	12	Adar>	5	6	7	8	9	10	11
13	14	15	16	17	18	19		12	13	14	15	16	17	18
20	21	22	23	24	25	26		19	20	21	22	23	24	25
27	28	1	2	3	4	5	<March	26	27	28	29	30	1	2
6	7	8	9	10	11	12	Adar 2>	3	4	5	6	7	8	9
13	14	15	16	17	18	19		10	11	12	13	14	15	16
20	21	22	23	24	25	26		17	18	19	20	21	22	23
27	28	29	30	31	1	2	<April	24	25	26	27	28	29	1
3	4	5	6	7	8	9	Nisan>	2	3	4	5	6	7	8
10	11	12	13	14	15	16		9	10	11	12	13	14	15
17	18	19	20	21	22	23		16	17	18	19	20	21	22
24	25	26	27	28	29	30		23	24	25	26	27	28	29
1	2	3	4	5	6	7	<May	30	1	2	3	4	5	6
8	9	10	11	12	13	14	Iyar>	7	8	9	10	11	12	13
15	16	17	18	19	20	21		14	15	16	17	18	19	20
22	23	24	25	26	27	28		21	22	23	24	25	26	27
29	30	31	1	2	3	4	<June	28	29	1	2	3	4	5
5	6	7	8	9	10	11	Sivan>	6	7	8	9	10	11	12
12	13	14	15	16	17	18		13	14	15	16	17	18	19
19	20	21	22	23	24	25		20	21	22	23	24	25	26
26	27	28	29	30	1	2	Tammuz>	27	28	29	30	1	2	3
3	4	5	6	7	8	9		4	5	6	7	8	9	10
10	11	12	13	14	15	16		11	12	13	14	15	16	17
17	18	19	20	21	22	23		18	19	20	21	22	23	24
24	25	26	27	28	29	30	Ab>	25	26	27	28	29	1	2
31	1	2	3	4	5	6	<August	3	4	5	6	7	8	9
7	8	9	10	11	12	13		10	11	12	13	14	15	16
14	15	16	17	18	19	20		17	18	19	20	21	22	23
21	22	23	24	25	26	27		24	25	26	27	28	29	30
28	29	30	31	1	2	3	Elul>	1	2	3	4	5	6	7
4	5	6	7	8	9	10	<Sept	8	9	10	11	12	13	14
11	12	13	14	15	16	17		15	16	17	18	19	20	21
18	19	20	21	22	23	24		22	23	24	25	26	27	28
25	26	27	28	29	30	1	Tishri>	29	1	2	3	4	5	6
2	3	4	5	6	7	8	<Oct	7	8	9	10	11	12	13
9	10	11	12	13	14	15		14	15	16	17	18	19	20
16	17	18	19	20	21	22		21	22	23	24	25	26	27
23	24	25	26	27	28	29	Heshvan>	28	29	30	1	2	3	4
30	31	1	2	3	4	5	<Nov	5	6	7	8	9	10	11
6	7	8	9	10	11	12		12	13	14	15	16	17	18
13	14	15	16	17	18	19		19	20	21	22	23	24	25
20	21	22	23	24	25	26		26	27	28	29	1	2	3
27	28	29	30	1	2	3	<Dec	4	5	6	7	8	9	10
4	5	6	7	8	9	10		11	12	13	14	15	16	17
11	12	13	14	15	16	17		18	19	20	21	22	23	24
18	19	20	21	22	23	24		25	26	27	28	29	1	2
25	26	27	28	29	30	31	Tebeth>	3	4	5	6	7	8	9

AD 562/Dionysius' Cycle II, Year 12/Metonic Year 9/Easter 09 April, Luna 19

S	*M*	*T*	*W*	*T*	*F*	*S*		*S*	*M*	*T*	*W*	*T*	*F*	*S*
1	2	3	4	5	6	7	Tebeth>	10	11	12	13	14	15	16
8	9	10	11	12	13	14	<Jan	17	18	19	20	21	22	23
15	16	17	18	19	20	21		24	25	26	27	28	29	1
22	23	24	25	26	27	28	Shevat>	2	3	4	5	6	7	8
29	30	31	1	2	3	4	<Feb	9	10	11	12	13	14	15
5	6	7	8	9	10	11		16	17	18	19	20	21	22
12	13	14	15	16	17	18		23	24	25	26	27	28	29
19	20	21	22	23	24	25	Adar>	30	1	2	3	4	5	6
26	27	28	1	2	3	4	<March	7	8	9	10	11	12	13
5	6	7	8	9	10	11		14	15	16	17	18	19	20
12	13	14	15	16	17	18		21	22	23	24	25	26	27
19	20	21	22	23	24	25	Nisan>	28	29	30	1	2	3	4
26	27	28	29	30	31	1	<April	5	6	7	8	9	10	11
2	3	4	5	6	7	8		12	13	14	15	16	17	18
9	10	11	12	13	14	15		19	20	21	22	23	24	25
16	17	18	19	20	21	22		26	27	28	29	30	1	2
23	24	25	26	27	28	29	Iyar>	3	4	5	6	7	8	9
30	1	2	3	4	5	6	<May	10	11	12	13	14	15	16
7	8	9	10	11	12	13		17	18	19	20	21	22	23
14	15	16	17	18	19	20		24	25	26	27	28	29	1
21	22	23	24	25	26	27	Sivan>	2	3	4	5	6	7	8
28	29	30	31	1	2	3	<June	9	10	11	12	13	14	15
4	5	6	7	8	9	10		16	17	18	19	20	21	22
11	12	13	14	15	16	17		23	24	25	26	27	28	29
18	19	20	21	22	23	24	Tammuz>	30	1	2	3	4	5	6
25	26	27	28	29	30	1		7	8	9	10	11	12	13
2	3	4	5	6	7	8	<July	14	15	16	17	18	19	20
9	10	11	12	13	14	15		21	22	23	24	25	26	27
16	17	18	19	20	21	22	Ab>	28	29	1	2	3	4	5
23	24	25	26	27	28	29		6	7	8	9	10	11	12
30	31	1	2	3	4	5	<August	13	14	15	16	17	18	19
6	7	8	9	10	11	12		20	21	22	23	24	25	26
13	14	15	16	17	18	19		27	28	29	30	1	2	3
20	21	22	23	24	25	26	Elul>	4	5	6	7	8	9	10
27	28	29	30	31	1	2	<Sept	11	12	13	14	15	16	17
3	4	5	6	7	8	9		18	19	20	21	22	23	24
10	11	12	13	14	15	16		25	26	27	28	29	1	2
17	18	19	20	21	22	23	Tishri>	3	4	5	6	7	8	9
24	25	26	27	28	29	30		10	11	12	13	14	15	16
1	2	3	4	5	6	7	<Oct	17	18	19	20	21	22	23
8	9	10	11	12	13	14		24	25	26	27	28	29	30
15	16	17	18	19	20	21	Heshvan>	1	2	3	4	5	6	7
22	23	24	25	26	27	28		8	9	10	11	12	13	14
29	30	31	1	2	3	4	<Nov	15	16	17	18	19	20	21
5	6	7	8	9	10	11		22	23	24	25	26	27	28
12	13	14	15	16	17	18		29	1	2	3	4	5	6
19	20	21	22	23	24	25		7	8	9	10	11	12	13
26	27	28	29	30	1	2		14	15	16	17	18	19	20
3	4	5	6	7	8	9	<Dec	21	22	23	24	25	26	27
10	11	12	13	14	15	16	Tebeth>	28	29	1	2	3	4	5
17	18	19	20	21	22	23		6	7	8	9	10	11	12
24	25	26	27	28	29	30		13	14	15	16	17	18	19
31								20						

AD 563/Dionysius Cycle II, Year 13/Metonic Year 10/Easter 25 March, Luna 15

S	*M*	*T*	*W*	*T*	*F*	*S*	*AD 563*	*S*	*M*	*T*	*W*	*T*	*F*	*S*
	1	2	3	4	5	6	Tebeth>		21	22	23	24	25	26
7	8	9	10	11	12	13	<Jan	27	28	29	1	2	3	4
14	15	16	17	18	19	20	Shebat>	5	6	7	8	9	10	11
21	22	23	24	25	26	27		12	13	14	15	16	17	18
28	29	30	31	1	2	3	<Feb	19	20	21	22	23	24	25
4	5	6	7	8	9	10	Adar>	26	27	28	29	30	1	2
11	12	13	14	15	16	17		3	4	5	6	7	8	9
18	19	20	21	22	23	24		10	11	12	13	14	15	16
25	26	27	28	1	2	3	<March	17	18	19	20	21	22	23
4	5	6	7	8	9	10		24	25	26	27	28	29	30
11	12	13	14	15	16	17	Nisan>	1	2	3	4	5	6	7
18	19	20	21	22	23	24		8	9	10	11	12	13	14
25	26	27	28	29	30	31		15	16	17	18	19	20	21
1	2	3	4	5	6	7	<April	22	23	24	25	26	27	28
8	9	10	11	12	13	14	Iyar>	29	30	1	2	3	4	5
15	16	17	18	19	20	21		6	7	8	9	10	11	12
22	23	24	25	26	27	28		13	14	15	16	17	18	19
29	30	1	2	3	4	5	<May	20	21	22	23	24	25	26
6	7	8	9	10	11	12	Sivan>	27	28	29	1	2	3	4
13	14	15	16	17	18	19		5	6	7	8	9	10	11
20	21	22	23	24	25	26		12	13	14	15	16	17	18
27	28	29	30	31	1	2	<June	19	20	21	22	23	24	25
3	4	5	6	7	8	9	Tammuz>	26	27	28	29	30	1	2
10	11	12	13	14	15	16		3	4	5	6	7	8	9
17	18	19	20	21	22	23		10	11	12	13	14	15	16
24	25	26	27	28	29	30		17	18	19	20	21	22	23
1	2	3	4	5	6	7	<July	24	25	26	27	28	29	1
8	9	10	11	12	13	14	Ab>	2	3	4	5	6	7	8
15	16	17	18	19	20	21		9	10	11	12	13	14	15
22	23	24	25	26	27	28		16	17	18	19	20	21	22
29	30	31	1	2	3	4	<August	23	24	25	26	27	28	29
5	6	7	8	9	10	11	Elul>	30	1	2	3	4	5	6
12	13	14	15	16	17	18		7	8	9	10	11	12	13
19	20	21	22	23	24	25		14	15	16	17	18	19	20
26	27	28	29	30	31	1	<Sept	21	22	23	24	25	26	27
2	3	4	5	6	7	8	Tishri>	28	29	1	2	3	4	5
9	10	11	12	13	14	15		6	7	8	9	10	11	12
16	17	18	19	20	21	22		13	14	15	16	17	18	19
23	24	25	26	27	28	29		20	21	22	23	24	25	26
30	1	2	3	4	5	6	<Oct	27	28	29	30	1	2	3
7	8	9	10	11	12	13	Heshvan>	4	5	6	7	8	9	10
14	15	16	17	18	19	20		11	12	13	14	15	16	17
21	22	23	24	25	26	27		18	19	20	21	22	23	24
28	29	30	31	1	2	3	Kislev>	25	26	27	28	29	30	1
4	5	6	7	8	9	10	<Nov	2	3	4	5	6	7	8
11	12	13	14	15	16	17		9	10	11	12	13	14	15
18	19	20	21	22	23	24		16	17	18	19	20	21	22
25	26	27	28	29	30	1	<Dec	23	24	25	26	27	28	29
2	3	4	5	6	7	8	Tebeth>	30	1	2	3	4	5	6
9	10	11	12	13	14	15		7	8	9	10	11	12	13
16	17	18	19	20	21	22		14	15	16	17	18	19	20
23	24	25	26	27	28	29		21	22	23	24	25	26	27
30	31							28	29					

AD 564/Dionysius' Cycle II, Year 14/Metonic Year 11 (emb.)/Easter 13 April, Luna 15

S	*M*	*T*	*W*	*T*	*F*	*S*		*S*	*M*	*T*	*W*	*T*	*F*	*S*
		1	2	3	4	5	Shebat>			1	2	3	4	5
6	7	8	9	10	11	12	<Jan	6	7	8	9	10	11	12
13	14	15	16	17	18	19		13	14	15	16	17	18	19
20	21	22	23	24	25	26		20	21	22	23	24	25	26
27	28	29	30	31	1	2	<Feb	27	28	29	30	1	2	3
3	4	5	6	7	8	9	Adar>	4	5	6	7	8	9	10
10	11	12	13	14	15	16		11	12	13	14	15	16	17
17	18	19	20	21	22	23		18	19	20	21	22	23	24
24	25	26	27	28	29	1	Adar2>	25	26	27	28	29	30	1
2	3	4	5	6	7	8	<March	2	3	4	5	6	7	8
9	10	11	12	13	14	15		9	10	11	12	13	14	15
16	17	18	19	20	21	22		16	17	18	19	20	21	22
23	24	25	26	27	28	29		23	24	25	26	27	28	29
30	31	1	2	3	4	5	<April	1	2	3	4	5	6	7
6	7	8	9	10	11	12	Nisan>	8	9	10	11	12	13	14
13	14	15	16	17	18	19		15	16	17	18	19	20	21
20	21	22	23	24	25	26		22	23	24	25	26	27	28
27	28	29	30	1	2	3	Iyar>	29	30	1	2	3	4	5
4	5	6	7	8	9	10	<May	6	7	8	9	10	11	12
11	12	13	14	15	16	17		13	14	15	16	17	18	19
18	19	20	21	22	23	24		20	21	22	23	24	25	26
25	26	27	28	29	30	31	Sivan>	27	28	29	1	2	3	4
1	2	3	4	5	6	7	<June	5	6	7	8	9	10	11
8	9	10	11	12	13	14		12	13	14	15	16	17	18
15	16	17	18	19	20	21		19	20	21	22	23	24	25
22	23	24	25	26	27	28	Tammuz>	26	27	28	29	30	1	2
29	30	1	2	3	4	5	<July	3	4	5	6	7	8	9
6	7	8	9	10	11	12		10	11	12	13	14	15	16
13	14	15	16	17	18	19		17	18	19	20	21	22	23
20	21	22	23	24	25	26	Ab>	24	25	26	27	28	29	1
27	28	29	30	31	1	2	<August	2	3	4	5	6	7	8
3	4	5	6	7	8	9		9	10	11	12	13	14	15
10	11	12	13	14	15	16		16	17	18	19	20	21	22
17	18	19	20	21	22	23		23	24	25	26	27	28	29
24	25	26	27	28	29	30	Elul>	30	1	2	3	4	5	6
31	1	2	3	4	5	6	<Sept	7	8	9	10	11	12	13
7	8	9	10	11	12	13		14	15	16	17	18	19	20
14	15	16	17	18	19	20		21	22	23	24	25	26	27
21	22	23	24	25	26	27	Tishri>	28	29	1	2	3	4	5
28	29	30	1	2	3	4	<Oct	6	7	8	9	10	11	12
5	6	7	8	9	10	11		13	14	15	16	17	18	19
12	13	14	15	16	17	18		20	21	22	23	24	25	26
19	20	21	22	23	24	25	Heshvan>	27	28	29	30	1	2	3
26	27	28	29	30	31	1	<Nov	4	5	6	7	8	9	10
2	3	4	5	6	7	8		11	12	13	14	15	16	17
9	10	11	12	13	14	15		18	19	20	21	22	23	24
16	17	18	19	20	21	22	Kislev>	25	26	27	28	29	1	2
23	24	25	26	27	28	29		3	4	5	6	7	8	9
30	1	2	3	4	5	6	<Dec	10	11	12	13	14	15	16
7	8	9	10	11	12	13		17	18	19	20	21	22	23
14	15	16	17	18	19	20		24	25	26	27	28	29	1
21	22	23	24	25	26	27	Tebeth>	2	3	4	5	6	7	8
28	29	30	31					9	10	11	12	13	14	15

AD 565/Dionysius' Cycle II, Year 15/Metonic Year 12/Easter 05 April, Luna 18

S	*M*	*T*	*W*	*T*	*F*	*S*		*S*	*M*	*T*	*W*	*T*	*F*	*S*
				1	2	3	Tebeth>	9	10	11	12	13	14	15
4	5	6	7	8	9	10	<Jan	16	17	18	19	20	21	22
11	12	13	14	15	16	17		23	24	25	26	27	28	29
18	19	20	21	22	23	24	Shebat>	1	2	3	4	5	6	7
25	26	27	28	29	30	31		8	9	10	11	12	13	14
1	2	3	4	5	6	7	<Feb	15	16	17	18	19	20	21
8	9	10	11	12	13	14		22	23	24	25	26	27	28
15	16	17	18	19	20	21	Adar>	29	30	1	2	3	4	5
22	23	24	25	26	27	28		6	7	8	9	10	11	12
1	2	3	4	5	6	7	<March	13	14	15	16	17	18	19
8	9	10	11	12	13	14		20	21	22	23	24	25	26
15	16	17	18	19	20	21	Nisan>	27	28	29	30	1	2	3
22	23	24	25	26	27	28		4	5	6	7	8	9	10
29	30	31	1	2	3	4	<April	11	12	13	14	15	16	17
5	6	7	8	9	10	11		18	19	20	21	22	23	24
12	13	14	15	16	17	18		25	26	27	28	29	30	1
19	20	21	22	23	24	25	Iyar>	2	3	4	5	6	7	8
26	27	28	29	30	1	2	<May	9	10	11	12	13	14	15
3	4	5	6	7	8	9		16	17	18	19	20	21	22
10	11	12	13	14	15	16		23	24	25	26	27	28	29
17	18	19	20	21	22	23	Sivan>	1	2	3	4	5	6	7
24	25	26	27	28	29	30		8	9	10	11	12	13	14
31	1	2	3	4	5	6	<June	15	16	17	18	19	20	21
7	8	9	10	11	12	13		22	23	24	25	26	27	28
14	15	16	17	18	19	20	Tammuz>	29	30	1	2	3	4	5
21	22	23	24	25	26	27		6	7	8	9	10	11	12
28	29	30	1	2	3	4	<July	13	14	15	16	17	18	19
5	6	7	8	9	10	11		20	21	22	23	24	25	26
12	13	14	15	16	17	18	Ab>	27	28	29	1	2	3	4
19	20	21	22	23	24	25		5	6	7	8	9	10	11
26	27	28	29	30	31	1		12	13	14	15	16	17	18
2	3	4	5	6	7	8	<August	19	20	21	22	23	24	25
9	10	11	12	13	14	15		26	27	28	29	30	1	2
16	17	18	19	20	21	22	Elul>	3	4	5	6	7	8	9
23	24	25	26	27	28	29		10	11	12	13	14	15	16
30	31	1	2	3	4	5	<Sept	17	18	19	20	21	22	23
6	7	8	9	10	11	12		24	25	26	27	28	29	1
13	14	15	16	17	18	19	Tishri>	2	3	4	5	6	7	8
20	21	22	23	24	25	26		9	10	11	12	13	14	15
27	28	29	30	1	2	3	<Oct	16	17	18	19	20	21	22
4	5	6	7	8	9	10		23	24	25	26	27	28	29
11	12	13	14	15	16	17	Heshvan>	30	1	2	3	4	5	6
18	19	20	21	22	23	24		7	8	9	10	11	12	13
25	26	27	28	29	30	31		14	15	16	17	18	19	20
1	2	3	4	5	6	7	<Nov	21	22	23	24	25	26	27
8	9	10	11	12	13	14	Kislev>	28	29	1	2	3	4	5
15	16	17	18	19	20	21		6	7	8	9	10	11	12
22	23	24	25	26	27	28		13	14	15	16	17	18	19
29	30	1	2	3	4	5	<Dec	20	21	22	23	24	25	26
6	7	8	9	10	11	12	Tebeth>	27	28	29	1	2	3	4
13	14	15	16	17	18	19		5	6	7	8	9	10	11
20	21	22	23	24	25	26		12	13	14	15	16	17	18
27	28	29	30	31				19	20	21	22	23		

AD 566/Dionysius' Cycle II, Year 16/Metonic Year 13/Easter 28 March, Luna 21

S	*M*	*T*	*W*	*T*	*F*	*S*		*S*	*M*	*T*	*W*	*T*	*F*	*S*
					1	2	Tebeth>						24	25
3	4	5	6	7	8	9	<Jan	26	27	28	29	1	2	3
10	11	12	13	14	15	16	Shebat>	4	5	6	7	8	9	10
17	18	19	20	21	22	23		11	12	13	14	15	16	17
24	25	26	27	28	29	30		18	19	20	21	22	23	24
31	1	2	3	4	5	6	<Feb	25	26	27	28	29	30	1
7	8	9	10	11	12	13	Adar>	2	3	4	5	6	7	8
14	15	16	17	18	19	20		9	10	11	12	13	14	15
21	22	23	24	25	26	27		16	17	18	19	20	21	22
28	1	2	3	4	5	6	<March	23	24	25	26	27	28	29
7	8	9	10	11	12	13	Nisan>	30	1	2	3	4	5	6
14	15	16	17	18	19	20		7	8	9	10	11	12	13
21	22	23	24	25	26	27		14	15	16	17	18	19	20
28	29	30	31	1	2	3	<April	21	22	23	24	25	26	27
4	5	6	7	8	9	10	Iyar>	28	29	30	1	2	3	4
11	12	13	14	15	16	17		5	6	7	8	9	10	11
18	19	20	21	22	23	24		12	13	14	15	16	17	18
25	26	27	28	29	30	1		19	20	21	22	23	24	25
2	3	4	5	6	7	8	<May	26	27	28	29	1	2	3
9	10	11	12	13	14	15	Sivan>	4	5	6	7	8	9	10
16	17	18	19	20	21	22		11	12	13	14	15	16	17
23	24	25	26	27	28	29		18	19	20	21	22	23	24
30	31	1	2	3	4	5	<June	25	26	27	28	29	30	1
6	7	8	9	10	11	12	Tammuz>	2	3	4	5	6	7	8
13	14	15	16	17	18	19		9	10	11	12	13	14	15
20	21	22	23	24	25	26		16	17	18	19	20	21	22
27	27	29	30	1	2	3	<July	23	24	25	26	27	28	29
4	5	6	7	8	9	10	Ab>	1	2	3	4	5	6	7
11	12	13	14	15	16	17		8	9	10	11	12	13	14
18	19	20	21	22	23	24		15	16	17	18	19	20	21
25	26	27	28	29	30	31	<August	22	23	24	25	26	27	28
1	2	3	4	5	6	7	Elul>	29	30	1	2	3	4	5
8	9	10	11	12	13	14		6	7	8	9	10	11	12
15	16	17	18	19	20	21		13	14	15	16	17	18	19
22	23	24	25	26	27	28		20	21	22	23	24	25	26
29	30	31	1	2	3	4	<Sept	27	28	29	1	2	3	4
5	6	7	8	9	10	11	Tishri>	5	6	7	8	9	10	11
12	13	14	15	16	17	18		12	13	14	15	16	17	18
19	20	21	22	23	24	25		19	20	21	22	23	24	25
26	27	28	29	30	1	2	Heshvan>	26	27	28	29	30	1	2
3	4	5	6	7	8	9	<Oct	3	4	5	6	7	8	9
10	11	12	13	14	15	16		10	11	12	13	14	15	16
17	18	19	20	21	22	23		17	18	19	20	21	22	23
24	25	26	27	28	29	30	Kislev>	24	25	26	27	28	29	1
31	1	2	3	4	5	6	<Nov	2	3	4	5	6	7	8
7	8	9	10	11	12	13		9	10	11	12	13	14	15
14	15	16	17	18	19	20		16	17	18	19	20	21	22
21	22	23	24	25	26	27		23	24	25	26	27	28	29
28	29	30	1	2	3	4	<Dec	30	1	2	3	4	5	6
5	6	7	8	9	10	11	Tebeth>	7	8	9	10	11	12	13
12	13	14	15	16	17	18		14	15	16	17	18	19	20
19	20	21	22	23	24	25		21	22	23	24	25	26	27
26	27	28	29	30	31		Shebat>	28	29	1	2	3	4	

AD 567/Dionysius' Cycle II, Year 17/Metonic Year 14 (emb.)/Easter 10 April, Luna 15

S	*M*	*T*	*W*	*T*	*F*	*S*		*S*	*M*	*T*	*W*	*T*	*F*	*S*
						1	Shebat>			1	2	3	4	5
2	3	4	5	6	7	8	<Jan	6	7	8	9	10	11	12
9	10	11	12	13	14	15		13	14	15	16	17	18	19
16	17	18	19	20	21	22		20	21	22	23	24	25	26
23	24	25	26	27	28	29	Adar>	27	28	29	30	1	2	3
30	31	1	2	3	4	5	<Feb	4	5	6	7	8	9	10
6	7	8	9	10	11	12		11	12	13	14	15	16	17
13	14	15	16	17	18	19		18	19	20	21	22	23	24
20	21	22	23	24	25	26	Adar 2>	25	26	27	28	29	30	1
27	28	1	2	3	4	5	<March	2	3	4	5	6	7	8
6	7	8	9	10	11	12		9	10	11	12	13	14	15
13	14	15	16	17	18	19		16	17	18	19	20	21	22
20	21	22	23	24	25	26		23	24	25	26	27	28	29
27	28	29	30	31	1	2	Nisan>	1	2	3	4	5	6	7
3	4	5	6	7	8	9	<April	8	9	10	11	12	13	14
10	11	12	13	14	15	16		15	16	17	18	19	20	21
17	18	19	20	21	22	23		22	23	24	25	26	27	28
24	25	26	27	28	29	30	Iyar>	29	30	1	2	3	4	5
1	2	3	4	5	6	7	<May	6	7	8	9	10	11	12
8	9	10	11	12	13	14		13	14	15	16	17	18	19
15	16	17	18	19	20	21		20	21	22	23	24	25	26
22	23	24	25	26	27	28	Sivan>	27	28	29	1	2	3	4
29	30	31	1	2	3	4	<June	5	6	7	8	9	10	11
5	6	7	8	9	10	11		12	13	14	15	16	17	18
12	13	14	15	16	17	18		19	20	21	22	23	24	25
19	20	21	22	23	24	25	Tammuz>	26	27	28	29	30	1	2
26	27	28	29	30	1	2	<July	3	4	5	6	7	8	9
3	4	5	6	7	8	9		10	11	12	13	14	15	16
10	11	12	13	14	15	16		17	18	19	20	21	22	23
17	18	19	20	21	22	23		24	25	26	27	28	29	1
24	25	26	27	28	29	30	Ab>	2	3	4	5	6	7	8
31	1	2	3	4	5	6	<August	9	10	11	12	13	14	15
7	8	9	10	11	12	13		16	17	18	19	20	21	22
14	15	16	17	18	19	20		23	24	25	26	27	28	29
21	22	23	24	25	26	27	Elul>	30	1	2	3	4	5	6
28	29	30	31	1	2	3	<Sept	7	8	9	10	11	12	13
4	5	6	7	8	9	10		14	15	16	17	18	19	20
11	12	13	14	15	16	17		21	22	23	24	25	26	27
18	19	20	21	22	23	24	Tishri>	28	29	1	2	3	4	5
25	26	27	28	29	30	1		6	7	8	9	10	11	12
2	3	4	5	6	7	8	<Oct	13	14	15	16	17	18	19
9	10	11	12	13	14	15		20	21	22	23	24	25	26
16	17	18	19	20	21	22	Heshvan>	27	28	29	30	1	2	3
23	24	25	26	27	28	29		4	5	6	7	8	9	10
30	31	1	2	3	4	5	<Nov	11	12	13	14	15	16	17
6	7	8	9	10	11	12		18	19	20	21	22	23	24
13	14	15	16	17	18	19	Kislev>	25	26	27	28	29	1	2
20	21	22	23	24	25	26		3	4	5	6	7	8	9
27	28	29	30	1	2	3	<Dec	10	11	12	13	14	15	16
4	5	6	7	8	9	10		17	18	19	20	21	22	23
11	12	13	14	15	16	17		24	25	26	27	28	29	30
18	19	20	21	22	23	24	Tebeth>	1	2	3	4	5	6	7
25	26	27	28	29	30	31		8	9	10	11	12	13	14

AD 568/Dionysius' Cycle II, Year 18/Metonic Year 15/ Easter 01 April, Luna 17

S	*M*	*T*	*W*	*T*	*F*	*S*		*S*	*M*	*T*	*W*	*T*	*F*	*S*
1	2	3	4	5	6	7	Tebeth>	15	16	17	18	19	20	21
8	9	10	11	12	13	14	<Jan	22	23	24	25	26	27	28
15	16	17	18	19	20	21	Shebat>	29	1	2	3	4	5	6
22	23	24	25	26	27	28		7	8	9	10	11	12	13
29	30	31	1	2	3	4	<Feb	14	15	16	17	18	19	20
5	6	7	8	9	10	11		21	22	23	24	25	26	27
12	13	14	15	16	17	18	Adar>	28	29	30	1	2	3	4
19	20	21	22	23	24	25		5	6	7	8	9	10	11
26	27	28	29	1	2	3	<March	12	13	14	15	16	17	18
4	5	6	7	8	9	10		19	20	21	22	23	24	25
11	12	13	14	15	16	17	Nisan>	26	27	28	29	30	1	2
18	19	20	21	22	23	24		3	4	5	6	7	8	9
25	26	27	28	29	30	31		10	11	12	13	14	15	16
1	2	3	4	5	6	7	<April	17	18	19	20	21	22	23
8	9	10	11	12	13	14		24	25	26	27	28	29	30
15	16	17	18	19	20	21	Iyar>	1	2	3	4	5	6	7
22	23	24	25	26	27	28		8	9	10	11	12	13	14
29	30	1	2	3	4	5	<May	15	16	17	18	19	20	21
6	7	8	9	10	11	12		22	23	24	25	26	27	28
13	14	15	16	17	18	19	Sivan>	29	1	2	3	4	5	6
20	21	22	23	24	25	26		7	8	9	10	11	12	13
27	28	29	30	31	1	2	<June	14	15	16	17	18	19	20
3	4	5	6	7	8	9		21	22	23	24	25	26	27
10	11	12	13	14	15	16	Tammuz>	28	29	30	1	2	3	4
17	18	19	20	21	22	23		5	6	7	8	9	10	11
24	25	26	27	28	29	30		12	13	14	15	16	17	18
1	2	3	4	5	6	7	<July	19	20	21	22	23	24	25
8	9	10	11	12	13	14	Ab>	26	27	28	29	1	2	3
15	16	17	18	19	20	21		4	5	6	7	8	9	10
22	23	24	25	26	27	28		11	12	13	14	15	16	17
29	30	31	1	2	3	4	<August	18	19	20	21	22	23	24
5	6	7	8	9	10	11		25	26	27	28	29	30	1
12	13	14	15	16	17	18	Elul>	2	3	4	5	6	7	8
19	20	21	22	23	24	25		9	10	11	12	13	14	15
26	27	28	29	30	31	1		16	17	18	19	20	21	22
2	3	4	5	6	7	8	<Sep	23	24	25	26	27	28	29
9	10	11	12	13	14	15	Tishri>	1	2	3	4	5	6	7
16	17	18	19	20	21	22		8	9	10	11	12	13	14
23	24	25	26	27	28	29		15	16	17	18	19	20	21
30	1	2	3	4	5	6	<Oct	22	23	24	25	26	27	28
7	8	9	10	11	12	13	Heshvan>	29	30	1	2	3	4	5
14	15	16	17	18	19	20		6	7	8	9	10	11	12
21	22	23	24	25	26	27		13	14	15	16	17	18	19
28	29	30	31	1	2	3	<Nov	20	21	22	23	24	25	26
4	5	6	7	8	9	10	Kislev>	27	28	29	1	2	3	4
11	12	13	14	15	16	17		5	6	7	8	9	10	11
18	19	20	21	22	23	24		12	13	14	15	16	17	18
25	26	27	28	29	30	1		19	20	21	22	23	24	25
2	3	4	5	6	7	8	<Dec	26	27	28	29	30	1	2
9	10	11	12	13	14	15	Tebeth>	3	4	5	6	7	8	9
16	17	18	19	20	21	22		10	11	12	13	14	15	16
23	24	25	26	27	28	29		17	18	19	20	21	22	23
30	31							24	25					

AD 569/Dionysius' Cycle II, Year 19/Metonic Year 16 (emb.)/Easter 21 April, Luna 18

S	*M*	*T*	*W*	*T*	*F*	*S*		*S*	*M*	*T*	*W*	*T*	*F*	*S*
		1	2	3	4	5	Tebeth>			26	27	28	29	1
6	7	8	9	10	11	12	Shebat>	2	3	4	5	6	7	8
13	14	15	16	17	18	19	<Jan	9	10	11	12	13	14	15
20	21	22	23	24	25	26		16	17	18	19	20	21	22
27	28	29	30	31	1	2	<Feb	23	24	25	26	27	28	29
3	4	5	6	7	8	9	Adar>	30	1	2	3	4	5	6
10	11	12	13	14	15	16		7	8	9	10	11	12	13
17	18	19	20	21	22	23		14	15	16	17	18	19	20
24	25	26	27	28	1	2	<March	21	22	23	24	25	26	27
3	4	5	6	7	8	9	Adar 2>	28	29	30	1	2	3	4
10	11	12	13	14	15	16		5	6	7	8	9	10	11
17	18	19	20	21	22	23		12	13	14	15	16	17	18
24	25	26	27	28	29	30		19	20	21	22	23	24	25
31	1	2	3	4	5	6	<April	26	27	28	29	1	2	3
7	8	9	10	11	12	13	Nisan>	4	5	6	7	8	9	10
14	15	16	17	18	19	20		11	12	13	14	15	16	17
21	22	23	24	25	26	27		18	19	20	21	22	23	24
28	29	30	1	2	3	4	<May	25	26	27	28	29	30	1
5	6	7	8	9	10	11	Iyar>	2	3	4	5	6	7	8
12	13	14	15	16	17	18		9	10	11	12	13	14	15
19	20	21	22	23	24	25		16	17	18	19	20	21	22
26	27	28	29	30	31	1	<June	23	24	25	26	27	28	29
2	3	4	5	6	7	8	Sivan>	1	2	3	4	5	6	7
9	10	11	12	13	14	15		8	9	10	11	12	13	14
16	17	18	19	20	21	22		15	16	17	18	19	20	21
23	24	25	26	27	28	29		22	23	24	25	26	27	28
30	1	2	3	4	5	6	<July	29	30	1	2	3	4	5
7	8	9	10	11	12	13	Tammuz>	6	7	8	9	10	11	12
14	15	16	17	18	19	20		13	14	15	16	17	18	19
21	22	23	24	25	26	27		20	21	22	23	24	25	26
28	29	30	31	1	2	3	<August	27	28	29	1	2	3	4
4	5	6	7	8	9	10	Ab>	5	6	7	8	9	10	11
11	12	13	14	15	16	17		12	13	14	15	16	17	18
18	19	20	21	22	23	24		19	20	21	22	23	24	25
25	26	27	28	29	30	31	Elul>	26	27	28	29	30	1	2
1	2	3	4	5	6	7	<Sept	3	4	5	6	7	8	9
8	9	10	11	12	13	14		10	11	12	13	14	15	16
15	16	17	18	19	20	21		17	18	19	20	21	22	23
22	23	24	25	26	27	28		24	25	26	27	28	29	1
29	30	1	2	3	4	5	Tishri>	2	3	4	5	6	7	8
6	7	8	9	10	11	12	<Oct	9	10	11	12	13	14	15
13	14	15	16	17	18	19		16	17	18	19	20	21	22
20	21	22	23	24	25	26		23	24	25	26	27	28	29
27	28	29	30	31	1	2	Heshvan>	30	1	2	3	4	5	6
3	4	5	6	7	8	9	<Nov	7	8	9	10	11	12	13
10	11	12	13	14	15	16		14	15	16	17	18	19	20
17	18	19	20	21	22	23		21	22	23	24	25	26	27
24	25	26	27	28	29	30	*Kislev>	28	29	1	2	3	4	5
1	2	3	4	5	6	7	<Dec	6	7	8	9	10	11	12
8	9	10	11	12	13	14		13	14	15	16	17	18	19
15	16	17	18	19	20	21		20	21	22	23	24	25	26
22	23	24	25	26	27	28	Tebeth>	27	28	1	2	3	4	5
29	30	31						6	7	8				

APPENDIX F

TIMELINE

5502 BC	Assumed date of creation by Hippolytus in his Canon and Chronicle.
5500 BC	Assumed date of creation by Sextus Julius Africanus in his Chronographiai.
4241 BC	Traditional date of beginning of Egyptian calendar.
4004 BC, 23 October	Approximate date of creation according to Archbishop James Ussher.
3761 BC	Approximate date of creation by the Hebrew calendar.
3110 BC, 21 December	Approximate beginning date of the Mayan calendar, slated to expire on 21 December 2012.
Ca. 2560 BC	Construction of the Great Pyramid of Khufu at Giza, Egypt.
Ca. 2500 BC	Construction of Stonehenge on Salisbury Plain, England.
753 BC	Traditional date of the founding of Rome.
509 BC	Roman landowners proclaim the Republic after their last king is expelled and the Senate is formalized.
505 BC	Cleisthenes initiates democratic government in Athens.

435 BC	Meton and his associates in Athens introduce the Metonic Cycle, a 19-year projection which synchronizes lunar and solar cycles.
45 BC, 1 January Roman Date 709 AUC	Rome begins using the Julian calendar.
12 BC	Following the advice of astronomers, Caesar Augustus proclaims a moratorium on leap years to correct the overage of three days caused by the inclusion of a leap year every third year rather than every fourth.
9 BC (29 February)	In what would otherwise properly be a leap year, the Romans do not include 29 February in order to correct their former practice of triennial leap years.
6 BC	Year of Jesus' birth according to German astronomer Johannes Kepler, who (incorrectly) asserted that the planetary alignment of Jupiter and Saturn in 7 BC created a nova which was viewed by the Magi and constituted the star of Bethlehem.
5 BC (29 February)	In what would otherwise properly be a leap year, the Romans again include no 29 February in order to correct their former practice of triennial leap years.
3 BC, 30 March, Hebrew Date 14 Nisan, 3758 Roman Date 751 AUC	Incarnation of God the Son in the womb of the Virgin Mary.
1 BC, 10 January	Full lunar eclipse visible in Palestine; best candidate for that later cited by Flavius

	Josephus as occurring shortly before the death of Herod the Great.
1 BC (29 February)	In what would otherwise properly be a leap year, the Romans once more include no 29 February in order to correct their former practice of triennial leap years.
1 BC, 25 March	Apparent date chosen by Dionysius Exiguus for the incarnation of God the Son into human flesh.
AD 4, 29 February	Romans once again begin counting leap years, this time quadrennially.
AD 31, 23 March, Friday Roman Date 784 AUC Hebrew Date 11 Nisan, 3791	Apparent date chosen by Dionysius Exiguus for the crucifixion of Jesus Christ in Jerusalem.
AD 31, 25 March, Sunday	Apparent date chosen by Dionysius Exiguus for the resurrection of Jesus Christ.
AD 33, 3 April, Hebrew Date 14 Nisan, 3793	Actual date of crucifixion of Jesus.
AD 33, 5 April, Hebrew Date 16 Nisan	Resurrection of Jesus Christ.
AD 33, 14 May, Hebrew Date 25 Iyyar	Jesus' ascension from the Mount of Olives.
Ca. AD 180	Irenaeus dates birth of Christ to forty-first year of Augustus.

Ca. AD 200	Constitutions of the Holy Apostles, a Christian document of unknown authorship, cites the birth of Christ as an annual Christian observance.
Ca. AD 202	Hippolytus in his Commentary on Daniel dates the birth of Christ as AM 5500.
Ca. AD 210	Tertullian dates birth of Christ to forty-first year of Augustus.
Ca. AD 210	Clement of Alexandria dates birth of Christ to twenty-eighth year of Augustus and 194 years before death of Commodus.
AD 221	Sextus Julius Africanus writes Chronographiai.
Ca. AD 236	Passover and Easter cycles are inscribed on the sarcophagus of Hippolytus.
AD 274	Emperor Aurelian constructs a temple to "Sol Invictus" (the Unconquerable Sun) in Rome.
AD 312	Following his vision of a cross, Constantine converts to Christianity. After his victory over rival emperor Maxentius, he becomes emperor over the Western portion of the Roman Empire. By his death he is sole emperor. His conversion begins Christianity's rise to political and social dominance in the Roman world.
AD 313	The Edict of Milan, issued by co-emperors Constantine and Licinius, legalizes Christianity and ends all forms of persecution.

AD 325	The Council of Nicaea dictates a Sunday celebration of Christ's resurrection, independent of the Hebrew calendar.
AD 329	Athanasius issues the first of his annual letters from Alexandria which establishes the date of Easter.
AD 354	The unknown author of the Chronograph of 354 cites the birth of Christ as 25 December.
AD 359	Rabbi Hillel II establishes modern pattern of intercalations for Hebrew calendar.
Ca. AD 415	Augustine of Hippo, On the Trinity, asserts 25 March as the date of both the incarnation and resurrection of Christ.
AD 457	Victorius constructs his 532-year cycle of Easter dates based on Roman lunar measurements.
AD 476	The city of Rome falls out of "Roman" (Byzantine) hands with the invasion of Ostrogothic king Odoacer and deposition of Romulus Augustulus.
AD 525	Dionysius Exiguus constructs his Liber de Paschate, or Paschal cycle. In the process, he proposes a new chronological system based on "Anno Domini," or year of the Lord.
AD 540	Byzantines re-gain control of western Mediterranean, establish capital at Ravenna, and construct Basilica of San Vitale. Stone Paschal calendar installed at Ravenna.
AD 632	Death of Muhammad ibn Abdullah, rise of Islam in Arabia.

AD 664	Synod of Whitby in England reaffirms the correctness of Dionysius' Paschal cycle.
Ca. AD 700	Construction of astronomical observation site by Quitu-cara directly on the Equator at the summit of Monte Catequilla in modern Ecuador.
AD 725	The Venerable Bede in England writes *De Temporum Ratione*, which extends Dionysius' dating system backward to include period Before Christ.
Ca. 1000	Construction of El Castillo temple by Maya/Toltecs at Chichen Itza, Mexico
1095	Pope Urban II calls for first Christian crusade to liberate Palestine from Muslim domination.
Ca. 1100	Construction of "Woodhenge III" time-keeping device by Mississippians at Cahokia Mounds, Illinois, USA.
AD 1282	In a letter to Pope Martin IV, Roger Bacon explains that the true vernal equinox has regressed by several days against the Julian calendar, and suggests a calendrical revision.
1405	Astronomical clock completed at Saint Mary's Church in Lübeck, Germany, complete with Paschal calendar wheel. First of many such astronomical clocks/paschal calendars in cities of the Hanseatic League.
1475	Meridiana installed by Paolo del Pozzo Toscanelli at Duomo in Florence.

1492	Christopher Columbus "discovers" the Western Hemisphere.
1517	Martin Luther nails his Ninety-Five Theses on the door of the castle church at Wittenberg, Germany, thus signaling the beginning of the Protestant Reformation.
1540	Construction of astronomical clock at Saint Paulus Dom in Münster, Germany. Includes 532-year calendar of (Julian) dates for Easter through 2071.
1572	Egnazio Danti installs astronomical quadrant on South-facing front wall of Santa Maria Novella in Florence.
1574	Danti installs armillary sphere on front wall of Santa Maria Novella in Florence.
1575	Danti partially installs meridiana inside Santa Maria Novella in Florence.
1575	Aloisius Lilius makes recommendations to a papal commission for a calendrical revision. They were later revised and supplemented by Christopher Clavius.
1576	Danti completes meridiana in Cathedral of San Petronio in Bologna.
1581	Danti completes meridian in papal chapel of Torre dei Venti at the Vatican.
1582, 24 February	Pope Gregory XIII issues Inter Gravissimus—a revision of the calendar which returned the true vernal equinox to 20–21 March and restored the accuracy of the paschal calendar.

1582, 5 October	According to the dictates of the new Gregorian calendar, 5–14 October are eliminated.
1655	Danti's meridiana at San Petronio in Bologna is lengthened by Giovanni Domenico Cassini.
1695	Cassini uses meridiana at Bologna to verify the need for eliminating the upcoming leap year day in 1700. His calculations also verify the Earth's elliptical orbit around the sun.
1700 (29 February)	In those nations following the Gregorian ("New Style") calendar, the first omission of 29 February is observed in a centennial year not divisible by 400.
1702	Francesco Bianchini constructs meridiana at Santa Maria degli Angeli et dei Martyri in Rome.
1752	Adoption of the Gregorian calendar in England and its dominions.
1918	Adoption of the Gregorian calendar for civil purposes in Russia.
1923	Adoption of the Gregorian calendar for civil purposes in Greece.

APPENDIX G

EASTER DATES 1956–2200[1]

Year	*Orthodox Easter Julian Calendar*	*Orthodox Easter Gregorian Calendar*	*Western Easter*
1956	23 April*	6 May*	1 April
1957	8 April**	21 April**	21 April**
1958	31 March	13 April	6 April
1959	20 April*	3 May*	29 March
1960	4 April	17 April	17 April*
1961	27 March	9 April	2 April
1962	16 April*	29 April*	22 April*
1963	1 April	14 April	14 April
1964	20 April*	3 May*	29 March
1965	12 April	25 April	18 April*
1966	28 March	10 April	10 April
1967	17 April*	30 April*	26 March
1968	8 April	21 April	14 April*
1969	31 March	13 April	6 April
1970	13 April*	26 April*	29 March
1971	5 April	18 April	11 April*
1972	27 March	9 April	2 April
1973	16 April*	29 April*	22 April*
1974	1 April	14 April	14 April
1975	21 April*	4 May*	30 March
1976	12 April**	25 April**	18 April**
1977	28 March	10 April	10 April
1978	17 April*	30 April*	26 March
1979	9 April	22 April	15 April*
1980	24 March	6 April	6 April
1981	13 April*	26 April*	19 April*
1982	5 April	18 April	11 April
1983	25 April*	8 May*	3 April

[1] Marcus Montes' "Calculation of the Ecclesiastical Calendar," http://www.smart.net/~mmontes/ec-cal.html, provides reliable lists of projected Easter dates through 2124. Tables and instructions for calculating future Easter dates for both Orthodox and Western ecclesiastical calendars, and a simple explanation of the mathematical formula, are found in *OCY* 809ff, 862ff.

* denotes an embolismic year.

** denotes the beginning of a Metonic cycle.

Year	***Orthodox Easter Julian Calendar***	***Orthodox Easter Gregorian Calendar***	***Western Easter***
1984	9 April	22 April	22 April*
1985	1 April	14 April	7 April
1986	21 April*	4 May*	30 March
1987	6 April	19 April	19 April*
1988	28 March	10 April	3 April
1989	17 April*	30 April*	26 March
1990	2 April	15 April	15 April*
1991	25 March	7 April	31 March
1992	13 April*	26 April*	19 April*
1993	5 April	18 April	11 April
1994	18 April*	1 May*	3 April
1995	10 April**	23 April**	16 April**
1996	1 April	14 April	7 April
1997	14 April*	27 April*	30 March
1998	6 April	19 April	12 April*
1999	29 March	11 April	4 April
2000	17 April*	30 April*	23 April*
2001	2 April	15 April	15 April
2002	22 April*	5 May*	31 March
2003	14 April	27 April	20 April*
2004	29 March	11 April	11 April
2005	18 April*	1 May*	27 March
2006	10 April	23 April	16 April*
2007	26 March	8 April	8 April
2008	14 April*	27 April*	23 March
2009	6 April	19 April	12 April*
2010	22 March	4 April	4 April
2011	11 April*	24 April*	24 April*
2012	2 April	15 April	8 April
2013	22 April*	5 May*	31 March
2014	7 April**	20 April**	20 April**
2015	30 March	12 April	5 April
2016	18 April*	1 May*	27 March
2017	3 April	16 April	16 April*
2018	26 March	8 April	1 April
2019	15 April*	28 April*	21 April*
2020	6 April	19 April	12 April
2021	19 April*	2 May*	4 April
2022	11 April	24 April	17 April*
2023	3 April	16 April	9 April
2024	22 April*	5 May*	31 March
2025	7 April	20 April	20 April*
2026	30 March	12 April	5 April
2027	19 April*	2 May*	28 March
2028	3 April	16 April	16 April*
2029	26 March	8 April	1 April

Year	*Orthodox Easter Julian Calendar*	*Orthodox Easter Gregorian Calendar*	*Western Easter*
2030	15 April*	28 April*	21 April*
2031	31 March	13 April	13 April
2032	19 April*	2 May*	26 March
2033	11 April**	24 April**	17 April**
2034	27 March	9 April	9 April
2035	16 April*	29 April*	25 March
2036	7 April	20 April	13 April*
2037	23 March	5 April	5 April
2038	12 April*	25 April*	25 April*
2039	4 April	17 April	10 April
2040	23 April*	6 May*	1 April
2041	8 April	21 April	21 April*
2042	31 March	13 April	6 April
2043	20 April*	3 May*	29 March
2044	11 April	24 April	17 April*
2045	27 March	9 April	9 April
2046	16 April*	29 April*	25 March
2047	8 April	21 April	14 April*
2048	23 March	5 April	5 April
2049	12 April*	25 April*	18 April*
2050	4 April	17 April	10 April
2051	24 April*	7 May*	2 April
2052	8 April**	21 April**	21 April**
2053	31 March	13 April	6 April
2054	20 April*	3 May*	29 March
2055	5 April	18 April	18 April*
2056	27 March	9 April	2 April
2057	16 April*	29 April*	22 April*
2058	1 April	14 April	14 April
2059	21 April*	4 May*	30 March
2060	12 April	25 April	18 April*
2061	28 March	10 April	10 April
2062	17 April*	30 April*	26 March
2063	9 April	22 April	15 April*
2064	31 March	13 April	6 April
2065	13 April*	26 April*	29 March
2066	5 April	18 April	11 April*
2067	28 March	10 April	3 April
2068	16 April*	29 April*	22 April*
2069	1 April	14 April	14 April
2070	21 April*	4 May*	30 March
2071	6 April**	19 April**	19 April**
2072	28 March	10 April	10 April
2073	17 April*	30 April*	26 March
2074	9 April	22 April	15 April*
2075	25 March	7 April	7 April

Year	*Orthodox Easter Julian Calendar*	*Orthodox Easter Gregorian Calendar*	*Western Easter*
2076	13 April*	26 April*	19 April*
2077	5 April	18 April	11 April
2078	25 April*	8 May*	3 April
2079	10 April	23 April	23 April*
2080	1 April	14 April	7 April
2081	21 April*	4 May*	30 March
2082	6 April	19 April	19 April*
2083	29 March	11 April	4 April
2084	17 April*	30 April*	26 March
2085	2 April	15 April	15 April*
2086	25 March	7 April	31 March
2087	14 April*	27 April*	20 April*
2088	5 April	18 April	11 April
2089	18 April*	1 May*	3 April
2090	10 April**	23 April**	16 April**
2091	26 March	8 April	8 April
2092	14 April*	27 April*	30 March
2093	6 April	19 April	12 April*
2094	29 March	11 April	4 April
2095	11 April*	24 April*	24 April*
2096	2 April	15 April	15 April
2097	22 April*	5 May*	31 March
2098	14 April	27 April	20 April*
2099	30 March	12 April	12 April
2100	18 April*	2 May*	28 March
2101	10 April	24 April	17 April*
2102	26 March	9 April	9 April
2103	15 April*	29 April*	25 March
2104	6 April	20 April	13 April*
2105	22 March	5 April	5 April
2106	11 April*	25 April*	18 April*
2107	3 April	17 April	10 April
2108	22 April*	6 May*	1 April
2109	7 April**	21 April**	21 April**
2110	30 March	13 April	6 April
2111	19 April*	3 May*	29 March
2112	3 April	17 April	17 April*
2113	26 March	9 April	2 April
2114	15 April*	29 April*	22 April*
2115	31 March	14 April	14 April
2116	19 April*	3 May*	29 March
2117	11 April	25 April	18 April*
2118	3 April	17 April	10 April
2119	16 April*	30 April*	26 March
2120	7 April	21 April	14 April*
2121	30 March	13 April	6 April

Year	*Orthodox Easter Julian Calendar*	*Orthodox Easter Gregorian Calendar*	*Western Easter*
2122	19 April*	3 May*	29 March
2123	4 April	18 April	11 April*
2124	26 March	9 April	2 April
2125	15 April*	29 April*	22 April*
2126	31 March	14 April	14 April
2127	20 April*	4 May*	30 March
2128	11 April**	25 April**	18 April**
2129	27 March	10 April	10 April
2130	16 April*	30 April*	26 March
2131	8 April	22 April	15 April*
2132	23 March	6 April	6 April
2133	12 April*	26 April*	19 April*
2134	4 April	18 April	11 April
2135	24 April*	8 May*	3 April
2136	8 April	22 April	22 April*
2137	31 March	14 April	7 April
2138	20 April*	4 May*	30 March
2139	5 April	19 April	19 April*
2140	27 March	10 April	3 April
2141	16 April*	30 April*	26 March
2142	8 April	22 April	15 April*
2143	24 March	7 April	31 March
2144	12 April*	26 April*	19 April*
2145	4 April	18 April	11 April
2146	24 April*	8 May*	3 April
2147	19 April**	23 April**	16 April**
2148	31 March	14 April	7 April
2149	20 April*	4 May*	30 March
2150	5 April	19 April	12 April*
2151	28 March	11 April	4 April
2152	16 April*	30 April*	23 April*
2153	1 April	15 April	15 April
2154	21 April*	5 May*	31 March
2155	13 April	27 April	20 April*
2156	28 March	11 April	11 April
2157	17 April*	1 May*	27 March
2158	9 April	23 April	16 April*
2159	25 March	8 April	8 April
2160	13 April*	27 April*	23 March
2161	5 April	19 April	12 April*
2162	28 March	11 April	4 April
2163	10 April*	24 April*	24 April*
2164	1 April	15 April	8 April
2165	21 April*	5 May*	31 March
2166	6 April**	20 April**	20 April**
2167	29 March	12 April	5 April

Year	*Orthodox Easter Julian Calendar*	*Orthodox Easter Gregorian Calendar*	*Western Easter*
2168	17 April*	1 May*	27 March
2169	9 April	23 April	16 April*
2170	25 March	8 April	1 April
2171	14 April*	28 April*	21 April*
2172	5 April	19 April	12 April
2173	25 April*	9 May*	4 April
2174	10 April	24 April	17 April*
2175	2 April	16 April	9 April
2176	21 April*	5 May*	31 March
2177	6 April	20 April	20 April*
2178	29 March	12 April	5 April
2179	18 April*	2 May*	28 March
2180	2 April	16 April	16 April*
2181	25 March	8 April	1 April
2182	14 April*	28 April*	21 April*
2183	30 March	13 April	13 April
2184	18 April*	4 May*	28 March
2185	10 April**	24 April**	17 April**
2186	26 March	9 April	9 April
2187	15 April*	29 April*	25 March
2188	6 April	20 April	13 April*
2189	29 March	12 April	5 April
2190	11 April*	25 April*	25 April*
2191	3 April	17 April	10 April
2192	22 April*	6 May*	1 April
2193	14 April	28 April	21 April*
2194	30 March	13 April	6 April
2195	19 April*	3 May*	29 March
2196	10 April	24 April	17 April*
2197	26 March	9 April	9 April
2198	15 April*	29 April*	25 March
2199	7 April	21 April	14 April*
2200	22 March	6 April	6 April

GLOSSARY OF TERMS

Ab Urbe Condita, or Anno Urbis Conditae (AUC). Latin for "from the founded city," or years since the founding of Rome in 753 BC.

Anno Domini (AD). Latin for "in the year of the Lord." Our current AD chronological understanding originated with the Paschal Cycle of Dionysius Exiguus, denoting the purported number of years since the birth of Jesus Christ.

Anno Hijrah (AH). The chronological count of the lunar-based Muslim calendar. Begun in AD 622 with the "hijrah," or flight of Muhammad and his followers from Mecca to Medina.

Ante Christum (AC). Latin for "before Christ." Usage of this designation began with the Venerable Bede in his *Ecclesiastical History* and *De Temporum Ratione* in about AD 725.

Before Common Era (BCE). Although BCE can also denote "Before the Christian Era," it is a modern designation equivalent to "Before Christ," or "BC" which is usually intended to avoid specific recognition of the life of Jesus Christ.

Book of Jubilees. A Jewish text of the intertestamental period (probably mid-second century BC), which is largely a reworking and interpretation of the material of Genesis. It was well known among ancient Jews and early Christians, but has never enjoyed wide recognition of canonical status.

Calends (Kalends). Latin term denoting the beginning or first day of a month; source of our term *calendar*.

Century. Any period of one hundred years in length. The usual calendrical designation of a century, according to our system derived from Dionysius Exiguus, is that it begins on 1 January in the year ending with "__01," and ends on 31 December in the year ending with "__00."

Common Era (CE). Although CE can also mean "Christian Era," it is a modern designation equivalent to "Anno Domini," or "AD"—usually intended to avoid specific recognition of the life of Jesus Christ.

Decade. Any period of ten years in length. The usual calendrical designation of a decade is that it begins on 1 January in a year ending with "1," and ends on 31 December of a year ending in "0."

Dionysian Era. Another term denoting the period known as "AD" or "CE"—the period designated by Dionysius Exiguus as beginning with the birth of Jesus Christ.

Easter Cycle (Paschal Cycle). Any of several cycles of varying lengths developed by Christians during the first several centuries of the Christian Era with the purpose of establishing the date of Easter decades into the future. Cycles of 112 or 84 years were developed by the early third century in Rome, and soon thereafter cycles of 95 years (five consecutive Metonic cycles) were developed in Alexandria.

Ecclesiastical Calendar. *Ekklesia* (εκκλησια) is Greek for "those called out," and has been used historically to mean "church," or the corporate assembly of Christians. The ecclesiastical calendar is the annual schedule of holidays and feasts when Christians celebrate events in the life of Jesus or the history of Christianity—for instance, Christmas and Easter.

Ecliptic. The Earth's tilt or inclination on its axis at nearly 23.5°. It is sometimes described as the intersection of the Earth's equatorial plane with its orbital plane. It is because of the ecliptic that our planet has noticeable seasons, especially in the temperate zones. The temperate zones are those regions of the planet which are "above" the Tropics in latitude (23.5° or more north or south of the Equator) and "below" the polar regions (66.5° north or south of the Equator).

Embolism. A calendrical device employed by many ancient civilizations for the purpose of synchronizing lunar and solar cycles. Since the lunar cycle of just over 29.5 days does not multiply conveniently into the solar cycle (year) of nearly 365.25 days, the embolism was the imposition of an additional lunar cycle of twenty-nine days (some say thirty days) seven times in every nineteen years. The result is most commonly known as the **Metonic Cycle.**

Enneakaidekaeterida. (ἐννεακαιδεκαετεριδα) Greek term meaning "nineteen-year cycle," commonly applied to the Metonic cycle.

Epact. A number used on some ancient calendars denoting the age of the moon on a particular calendar date. On the Paschal Cycle of Dionysius Exiguus, the epact was noted for 22 March, the day assumed to be directly after the vernal equinox.

Equinox. Latin term meaning "equal night," which also implies equal daytime. It denotes those two astronomical points in the year at which the sun shines directly over the Earth's Equator, at roughly 21 March and 21 September. The season following the March equinox is spring in the Northern Hemisphere, when the daylight hours stretch beyond twelve hours per day and the hours of darkness shrink below twelve hours. Conversely, the season following the September equinox is autumn in the Northern Hemisphere, when the daylight hours sink below twelve hours per day and the hours of darkness expand accordingly. In the Southern Hemisphere, the effect is the opposite, with spring beginning at the September equinox and autumn at the March equinox.

Gregorian Calendar A revision of the Julian calendar accomplished under the administration of Pope Gregory XIII and instituted in 1582. The small inaccuracy (eleven minutes and fifteen seconds) in the Julian calendar's assumed length of the year (365.25 days) was corrected by eliminating ten days from October 1582. Moreover, in order to prevent the error from creeping back into the calendar it was declared that the centennial years (those ending in "__00") shall not be counted as leap years unless they are divisible by 400, not just by four as in other leap years.

Ides. Latin term originally denoting the date of the full moon. When the Julian calendar was instituted and the length of months was divorced from the length of the lunar cycle, the ides was designated as the thirteenth day of months having thirty days, and the fifteenth day of months having thirty-one days.

Integral Age. A belief stemming from the intertestamental period, and popular among some ancient Jews and early Christians, that true prophets sometimes were born and died on the same calendar day—thus signifying divine approval. Some have therefore postulated that the incarnation of Jesus occurred on the Julian vernal equinox (25 March), and that his crucifixion likewise occurred 30+ years later on 25 March (or in some cases that his resurrection from the dead occurred on this date). Alternatively, from the standpoint of the Hebrew calendar, it is equally possible to postulate that his conception and death both occurred on the feast of Passover.

Inquisition. Any one of several institutions devised by the Roman Catholic Church in the interest of prosecuting heretical practices. Begun in the late twelfth century, the Inquisition in some places became notorious for its use of torture to extract confessions from accused heretics. Contrary to popular belief, the Inquisition was rarely concerned with practices of Jews or Muslims, and in almost every case turned over convicted criminals to secular authorities for punishment.

Julian Calendar. A Roman calendrical revision accomplished under the administration of Julius Caesar and instituted in 45 BC. A great improvement upon the older Roman calendar which included only ten months and 304 days, it nevertheless contained a small error in counting every fourth year as a leap year. Its small error was corrected in the Gregorian calendrical revision of AD 1582.

Kalends. See "Calends."

Luna. Latin for "moon."

Lunar Cycle. The measure of the length of time it takes for the moon to pass through all of its phases. As of AD 2000, a lunar cycle is 29 days, 12 hours, 44 minutes, and nearly 3 seconds in length. It is the basis for the concept of "month," although since the institution of the Julian Calendar the lunar cycle itself has been officially ignored or divorced from the months of the calendar.

Metonic Cycle. A cycle of nineteen years developed by Meton and his associates in Athens in the fifth century BC. It was discovered that this was the closest possible synchronization of lunar and solar cycles: Nineteen solar years is within a few hours of equaling 235 lunar cycles. In order to accomplish this synchronization, the Metonic cycle used by numerous ancient societies included a thirteenth lunar cycle ("embolism") in seven out of every nineteen years—most commonly in years 3, 5, 8, 11, 13, 16, and 19.

Millennium. Any period of one thousand years in length. The usual calendrical designation of a millennium, according to the system derived from Dionysius Exiguus, is that it begins on 1 January in the year ending with "_001," and ends on 31 December in the year ending with "_000." For instance, the current millennium and the twenty-first century of the modern era began on 1 January, 2001, whereas the previous millennium and the twentieth century ended on 31 December, 2000.

New Style Calendar. See "Gregorian Calendar."

Nones. Latin designation for the first quarter of the lunar cycle. In the Julian calendar revision of 45 BC, which divorced the months of the calendar from the lunar cycle, it came to designate the seventh day of the month in those months with thirty-one days, and the fifth day of the month in those months with thirty days.

Old Style Calendar. See "Julian Calendar."

Paschal Cycle. See "Easter Cycle."

Saturnalia. An ancient Roman festival held in honor of Saturn, the father of some other Greco-Roman deities. Its earliest recognition dates from the late third century BC, and was observed 17–23 December. Although it appears to have been fairly popular, with its temporary reversal of social roles between masters and servants, there is no evidence that it achieved any kind of official status until the construction of the Sol Invictus temple in Rome under the rule of Aurelian (AD 274).

Sidereal Year. The measure of the length of time it takes for the earth to make one full revolution around the sun. As of AD 2000, an average Sidereal Year is 365 days, 6 hours, 9 minutes, and 9 seconds in length, which is 20 minutes and 24 seconds longer than the **Tropical Year**. As the Earth slows in its orbit around the Sun, the Sidereal Year is lengthening by about one second every 2,000 years.

Solar Cycle. Any measure of the Earth's annual circuit around the sun, roughly equal to a "year." The most common definition of a year for calendrical purposes is the length of time from 1 January of a given year until 1 January of the following year. The most common astronomical definition of a year is the **Tropical Year** (see below), which is the length of time from the vernal equinox of one year to the vernal equinox of the next year. The true definition of the length of time taken by the Earth for a complete circuit around the sun is the **Sidereal Year** (see above), which is slightly longer than the Tropical Year.

Solis. Latin for "sun."

Solstice. Latin term for "sun standing fast," denoting those two astronomical points in the year when the sun reaches its apparent highest point in the northern sky (approximately 21 June) and the southern sky (approximately 21 December). This is not due to any actual movement on the part of the sun or Earth, but the Earth's inclination at nearly 23.5° on its axis. The June solstice is the beginning of summer in the Northern Hemisphere, and the hours of daylight begin to shorten thereafter. The December solstice is the beginning of winter in the Northern Hemisphere, and the hours of daylight begin to lengthen thereafter. In the Southern Hemisphere, the effect is the opposite, with winter beginning at the June solstice and summer beginning at the December solstice.

Tropical Year. The measure of the length of time from the vernal (spring) equinox of one year to the vernal equinox of the next year. As of AD 2000, an average Tropical Year is 365 days, 5 hours, 48 minutes, and 45 seconds in length, which makes it 20 minutes and 24 seconds shorter than the **Sidereal Year**. Many ancient societies celebrated the vernal equinox as the beginning of a new year.

Vernum. Latin for "spring."

World Week. A mystical historical understanding which was common among ancient Jews and early Christians. Stemming from at least the second century BC and the ***Book of Jubilees***, the World Week is the assertion that human history is analogous to the seven days of creation in Genesis chapter one. Furthermore, it asserts that just as God completed creation within six days, his plan for human history will be completed within the course of 6,000 years.

BIBLIOGRAPHY

I. PRIMARY WORKS

Allen, Pauline, and Wendy Mayer. *Early Church Fathers: John Chrysostom.* Florence, KY: Routledge, 1999.

Bede. *Bede: The Reckoning of Time* Wallis, Faith, trans. Liverpool University Press-Translated Texts for Historians. Liverpool: Liverpool University Press, 1999.

Galilei, Galileo. *Opere.* National Edition. Favaro, Antonio, ed. Florence: Barbèra, 1890–1909. http://pinakes.imiss.fi.it:8080 pinakestext/annotate/seam?conversationId=6130 (accessed 15 November 2012).

Kepler, Johannes. *Harmonices Mundi.* Linz: Godofredi Tampachii, 1619. http://archive.org/http://archive.org/stream/ioanniskepplerih00kepl#page/n10/mode/1up (accessed 27 November 2012).

________. *Astronomia Nova.* Prague: P. Ramus, 1609. http://archive.org/stream/Astronomianovaa00Kepl#page/n27/mode/1up (accessed 1 December 2012).

Migne, J. P., et. al., eds. *Patrologia Graeca. Patrologiae cursus completus: series Graeca.* 162 vols. Paris: 1857–1867.

________. *Patrologia Latina. Patrologiae cursus completus: series Latina.* 221 vols. Paris: 1844–1902.

Monumenta Germaniae Historica. http://www.dmgh.de (accessed 8 November 2011).

Mynors, R. A. B., ed. *Cassiodoris Senatoris Institutiones.* Oxford: The Clarendon Press, 1937.

Pearse, Roger. "The Chronography of 354 AD. Part 12: Commemorations of the Martyrs." *MGH Chronica Minora* I (1892). http://www.tertullian.org/fathers/chronography_of_354_12depositions_martyrs.htm (accessed 8 November 2011).

________. "St. Jerome, Chronicle (2004–5)." The Tertullian Project. http://www.tertullian.org/fathers/jerome_chronicle _03_part2.htm (accessed 6 November 2011).

Pelikan, Jaroslav, ed. *Luther's Works.* American Edition. 55 vols. Jaroslav Pelikan, ed. Saint Louis: Concordia, 1958.

Roberts, Alexander, and James Donaldson, eds. *The Ante-Nicene Fathers.* Grand Rapids: William B. Eerdmans, 1989 [1885].

Schaff, Philip, ed. *The Nicene and Post-Nicene Fathers of the Christian Church.* 2 series; 28 vols. Grand Rapids: William B. Eerdmans, 1991 [1890].

Schmidt, Thomas C. "The Chronicon of Hippolytus." 2010. http://www.chronicon.net/chroniconfiles/Hippolytus%20and%20 December%2025th.pdf (accessed 11 November 2011).

Whitby, Michael, and Mary Whitby, trans. and eds. *Chronicon Paschale, 284–628 AD.* Liverpool: Liverpool University Press, 1989.

II. Secondary Works

Acts of the Fifth International Congress of Coptic Studies, 3 vols. Washington: International Association of Coptic Studies, 1992.

Adler, William. *Time Immemorial: Archaic History and Its Sources in Christian Chronography from Julius Africanus to George Syncellus.* Washington: Dumbarton Oaks, 1989.

Aveni, Anthony F. *Empires of Time: Calendars, Clocks, and Cultures.* New York: Basic Books, 1989.

Bacchiocchia, Samuele. *Anti-Judaism and the Origin of Sunday.* Rome: Pontifical Gregorian University Press, 1975.

Bardenhewer, Otto. *Patrology: The Lives and Works of the Fathers of the Church.* St. Louis: B. Herder, 1908.

Barnes, Timothy David. "The Date of Herod's Death," *JTS* 19 (1968): 204–219.

Barr, James. *Biblical Chronology: Legend or Science?* The Ethel M. Wood Lecture 1987. Delivered at the Senate House, University of London. London: University of London, 1987.

________. "Why the World was Created in 4004 BC: Archbishop Ussher and Biblical Chronology." *Bulletin of the John Rylands University Library* 67, 2 (Spring 1985): 575–608.

Beckwith, Roger T. "Saint Luke, the Date of Christmas and the Priestly Courses at Qumran." *Revue de Qumran* 9, no. 1 (1977): 73–94.

Behe, Michael. *Darwin's Black Box: The Biochemical Challenge to Evolution.* New York: Simon and Schuster, 2006.

Behr, John. *The Mystery of Christ: Life in Death.* Yonkers, NY: Saint Vladimir's Seminary Press, 2006

Bernegger, P. M. "The Affirmation of Herod's Death in 4 BC," *JTS* 34 (1983): 526–531.

Bible Prophecy Year of 360 Days. http://www.360calendar.com/bible-prophecy-360-days-calendar-2a.htm (accessed 11 November 2011).

Blackburn, Bonnie, and Leofranc Holford-Strevens. *The Oxford Companion to the Year: An Exploration of Calendar Customs and Time-Reckoning.* Oxford: Oxford University Press, 1999.

Bliss, S. *Analysis of Sacred Chronology, with the Elements of Chronology; and the Numbers of the Hebrew Text Vindicated.* Oakland, CA: Pacific Press, 1887.

Bònoli, Febrizio. "1655–2005: 350 Years of the Great Meridian Line by G. D. Cassini in the Basilica of San Petronio in Bologna," http://www.bo.astro.it/universo/cassini/meridian_ing.htm.

Booker, Bruce. "The Feasts of the Lord (Moedim) and their Fulfillment in Messiah Yeshua," (2008) http://www.scribd.com/doc/ 62705266/The-Moedim-of-Yeshua-The-Feasts-of-the-Lord-by-Bruce-Booker.

Borst, Arno. *Die Karolingische Kalenderreform. Monumenta Germaniae Historica Schriften, Band 46.* Hanover: Hahnsche Buchhandlung, 1998.

Brent, Allen. *Hippolytus and the Roman Church in the Third Century: Communities in Tension before the Emergence of a Monarch-Bishop.* Leiden: E. J. Brill, 1995.

Brown, Farrell. "Why Does Easter's Date Wander?" *Christian History and Biography,* April 9, 2004. http://www.christianitytoday.com/history/newsletter/2004/apr9a.html (accessed 17 June 2009).

Bruce, Frederick Fyvie. *Jesus and Christian Origins Outside the New Testament.* Grand Rapids: Eerdmans, 1974.

Burgess, Stanley M., ed. *The New International Dictionary of Pentecostal and Charismatic Movements.* Grand Rapids: Zondervan, 2002.

Cahokia Mounds State Historic Site. http://cahokiamounds.org (accessed 23 November 2011).

Cantalamessa, Raniero. *Ostern in der Alten Kirche.* Bern: Peter Lang, 1981.

Carpenter, Joel A. *Revive Us Again: The Reawakening of American Fundamentalism.* New York: Oxford, 1999.

Carson, Donald A., ed. *From Sabbath to Lord's Day: A Biblical, Historical, and Theological Investigation.* Grand Rapids: Zondervan, 1982.

Caspar, Max. *Kepler*. New York: Dover, 1993.

The Catholic Encyclopedia. 2009. http://www.newadvent.org/cathen (accessed 21 May 2009).

Cheney, Christopher Robert, and Michael Jones, eds. *A Handbook of Dates for Students of English History.* 2nd ed. Cambridge: University Press, 2000.

Chichen Itza-Architecture. http://www.world-mysteries.com/chichen_architecture.htm (accessed 30 December 2011).

Clouse, Robert G., Robert N. Hosack, and Richard V. Pierard. *The New Millennium Manual: A Once and Future Guide.* Grand Rapids: Baker, 1999.

Columbia Electronic Encyclopedia, 6th ed. "Development of the Modern Calendar." 2007. http://www.infoplease.com/ce6/society/A0857113.html (accessed 1 June 2009).

Conservapedia: The Trustworthy Encyclopedia. "Luke and the Census." http://www.conservapedia.com/Luke_and_the _Census (accessed 11 November 2011).

Conybeare, F. C. "The History of Christmas." *The American Journal of Theology* III, No. 1 (January 1899): 1–21.

Cornwell, Jim A. "The Star of Bethlehem and Coma Berenices." http://www.mazzaroth.com/ChapterSix/TheStarOfBethlehem.htm (accessed 19 December 2010).

Coveney, Peter, and Roger Highfield. *The Arrow of Time: A Voyage Through Science to Solve Time's Greatest Mystery.* New York: Fawcett Columbine, 1990.

Cross, F. L., ed. *The Oxford Dictionary of the Christian Church.* Oxford: University Press, 1974.

Crossan, John Dominic, and Richard G. Watts. *Who is Jesus?: Answers to Your Questions about the Historical Jesus.* Louisville: Westminster John Knox Press, 1999.

Croucher, Rowland. John Mark Ministries. "Famous Predictions of Jesus' Return." http://jmm.aaa.net.au/articles/14229.htm (accessed 31 December 2011).

Davies, Paul. *The Mind of God: The Scientific Basis for a Rational World.* New York: Simon and Schuster, 1993.

Dawkins, Richard. *The Blind Watchmaker: Why the Evidence of Evolution Reveals a Universe Without Design.* New York: W. W. Norton, 1996.

Declercq, Georges. *Anno Domini: The Origins of the Christian Era.* Brussels: Brepols, 2000.

Dembski, William A. *Intelligent Design: The Bridge Between Science and Theology.* Downers Grove, IL: IVP Academic, 1999.

Dix, Dom Gregory. *The Shape of the Liturgy.* New York: Continuum, 2005 [1978].

Dolan, Graham. *The Greenwich Guide to Measuring Time.* Greenwich: Royal Observatory Greenwich, 2001.

Dohrn-van Rossum, Gerhard. *History of the Hour: Clocks and Modern Temporal Orders.* Chicago: University of Chicago Press, 1996.

Draper, John William. *History of the Conflict Between Religion and Science.* Cambridge: University Press, 2009 [1874].

Duncan, David Ewing. *The Calendar: Humanity's Epic Struggle to Determine a True and Accurate Year.* London: Fourth Estate, 1999.

Edson, Evelyn. "World Maps and Easter Tables: Medieval Maps in Context." *Imago Mundi* 48 (1996): 25–42.

Encyclopedia Judaica. 18 vols. Cecil Roth, ed. Jerusalem: Keter/Macmillan, 1978.

Family Radio. 2011. "Judgment Day." http://www.familyradio.com/graphical/literature/judgment/judgment.html (accessed 19 May 2011; discontinued).

Famous Trials in American History. "Tennessee vs. John Scopes: The 'Monkey Trial,' 1925." http://www.law.umkc.edu/faculty/projects/ftrials/scopes/scopes.htm (accessed 31 December 2011).

Feeney, Denis. *Caesar's Calendar: Ancient Time and the Beginnings of History.* Berkeley: University of California Press, 2007.

Filkin, David. *Stephen Hawking's Universe: The Cosmos Explained.* New York: Basic Books, 1997.

Finegan, Jack. *Handbook of Biblical Chronology: Principles of Time Reckoning in the Ancient World and Problems of Chronology in the Bible.* Revised edition. Peabody, MA: Hendrickson, 1998.

Finocchiaro, Maurice A. *Retrying Galileo, 1633–1992.* Berkeley: University of California Press, 2005.

Finocchiaro, Maurice A., ed. *The Galileo Affair: A Documentary History.* Berkeley: University of California Press, 1989.

Frank, Tenney, ed. *An Economic Survey of Ancient Rome.* New York: Pageant Books, 1959.

Freeman, Charles. *The Closing of the Western Mind: The Rise of Faith and the Fall of Reason.* New York: Vintage, 2005.

Frend, W. H. C. *Martyrdom and Persecution in the Early Church.* Grand Rapids: Baker, 1965.

Galison, Peter. *Einstein's Clocks, Poincaré's Maps: Empires of Time.* New York: W. W. Norton, 2003.

Gateways to Babylon. "Akitu: The Babylonian New Year's Festival." http://www.gatewaystobabylon.com/religion /akitu.htm (accessed 11 November 2011).

Ginzel, F. K. *Handbuch der Mathematischen und Technischen Chronologie*. Leipzig: J. C. Hinrichs, 1906–1914.

Gertz, Steven. "Upon Further Review . . . : 360 Years Later, the Pope Admits the Church's 'Error of Judgment' toward Galileo." *Christian History* 76 (Vol. XXI, No. 4; 2002), 17.

Goetz, David. "The Monkey Trial." *Christian History* 55 (July 1997), http://www.christianitytoday.com/ch/1997/issue55/55h010.html (accessed 1 September 2009).

Gould, Stephen Jay. "Dousing Diminutive Dennis's Debate." *Natural History* 103, No. 4 (April 1994): 4–8.

Grafton, Anthony T. *Joseph Scaliger: A Study in the History of Classical Scholarship.* 2 vols. New York: Oxford, 1983.

________. "Joseph Justus Scaliger and Historical Chronology: The Rise and Fall of a Discipline." *History and Theory* 14, No. 2 (May 1975): 156–185.

Grant, Michael. "The Bimillennium." *Greece & Rome*, Second Series, Vol. 47, No. 1 (April 2000): 1–2.

Grout, James. Encyclopaedia Romana. "Sol Invictus and Christ-mas." http://penelope.uchicago.edu/~grout/encyclopaedia_romana/calendar/invictus.html (accessed 21 November 2011).

Grudem, Wayne A. *Bible Doctrine: Essential Teachings of the Christian Faith.* Grand Rapids: Zondervan, 1999.

Hanna, Nick. *The Millennium: A Rough Guide to the Year 2000.* London: Penguin Books, 1999.

Hannah, Robert. *Greek and Roman Calendars: Constructions of Time in the Classical World.* London: Duckworth, 2005.

Hawking, Stephen. *A Brief History of Time.* New York: Bantam Books, 1988.

Hebcal Jewish Calendar. http://www.hebcal.com/converter (accessed 29 December 2011).

Heilbron, J. L. *The Sun in the Church: Cathedrals as Solar Observatories.* Cambridge: Harvard University Press, 1999.

The History of Christmas. http://www.holidays.net/christmas /story.htm (accessed 30 August 2010).

Hoehner, Harold W. *Chronological Aspects of the Life of Christ.* Grand Rapids: Zondervan, 1977.

Holford-Strevens, Leofranc. *The History of Time: A Very Short Introduction.* New York: Oxford, 2005.

Humerfelt, Sigurd. GPS Enthusiast's Web Ring. "Earth according to WGS 84." http://home.online.no/~sigurdhu/Grid_1deg.htm (accessed 31 December 2011).

Jaritz, Gerhard, and Gerson Moreno-Riano, eds. *Time and Eternity: The Medieval Discourse*. Brepols: Turnhout, 2003.

Jaubert, Annie. *La Date de la Céne: calendrier biblique et liturgie chretienne.* Etudes biblique. Paris: Gabalda, 1957.

Jeffrey, Grant R. "The 360 Day Prophetic Year of the Bible." Trinity Consulting. 2010. http://xwalk.ca/360day.html (accessed 12 December 2010).

________. *Armageddon: Appointment with Destiny.* Colorado Springs: WaterBrook Press, 1997.

________. *The New Temple and the Second Coming: The Prophecy that Points to Christ's Return in Your Generation.* WaterBrook, 2007.

________. *Countdown to the Apocalypse: Learn to Read the signs that the Last Days Have Begun.* WaterBrook, 2008.

Jeremias, Joachim. *The Eucharistic Words of Jesus.* New York: Scribner, 1955.

Jones, Charles W. "The Victorian and Dionysiac Paschal Tables in the West." *Speculum* 9, No. 4 (October 1934): 408–421.

Kendrick, Graham. Graham Kendrick: The Official Website. “Deck the Halls.” http://www.grahamkendrick.co.uk/insight/teaching/alpha _deck.php (accessed 30 September 2010).

Kilkenny, Niall. “The Saint Bartholomew’s Day Massacre,” http://www.reformation.org/saint-barts-day-massacre.html (accessed 31 December 2011).

Killian, Bruce A. “Dionysius Exiguus Got It Right.” 2002. http://www.scripturescholar.com/DionysiusExiguus.htm (accessed 11 November 2011).

Killian, Greg. The Watchman. “Passover Chronology.” http://www.betemunah.org/chronology.htm.

Kirsch, Jonathan. *The Grand Inquisitor’s Manual: A History of Terror in the Name of God*. New York: HarperOne, 2009.

Koenig, Don. The Prophetic Years: Bible Prophecy and Conservative Christian Commentary. “When Will Jesus Christ Return?: Projecting a Fairly Realistic Date for the Second Coming of Jesus Christ.” http://www.thepropheticyears.com/comments/countdown_to_second_coming.htm (accessed 31 December 2011).

Krupp, E. C. “The Great 2012 Scare: Your 2012 Defense Kit.” *Sky and Telescope* (November 2009): 22–26.

Krusch, Bruno. *Abhandlungen der Preußischen Akademie der Wissenschaften. Philosophisch-historische Klasse, Nr. 8: Studien zur christlich-mittelalterlichen Chronologie.* Berlin: de Gruyter, 1937.

Kuzenkov, Pavel. “How old is the world? The Byzantine era κατα Ρωμαιους and its rivals.” *21st International Congress of Byzantine Studies* (London, 2006).

LaHaye, Tim, and Jerry B. Jenkins. *Left Behind: A Novel of the Earth’s Last Days.* Carol Stream, IL: Tyndale House, 1995.

Landes, Richard. “The Fear of an Apocalyptic Year 1000: Augustinian Historiography, Medieval and Modern.” *Speculum* 75, No. 1 (January 2000): 97–145.

Larkin, Clarence. *Dispensational Truth, or God’s Plan and Purpose in the Ages.* Santa Fe: Sun Publishing, 1998 [1918].

Larson, Frederick A. The Star of Bethlehem. "The Starry Dance." http://www.bethlehemstar.net/dance/dance.htm (accessed 11 November 2011).

Lindberg, Carter. *The European Reformations.* New York: Wiley-Blackwell, 1996.

Lindberg, David, and Ronald Numbers, eds. *God and Nature: Historical Essays on the Encounter Between Christianity and Science.* Berkeley: University of California Press, 1986.

Lindsey, Hal. *The Late Great Planet Earth.* Grand Rapids: Zondervan, 1998 [1970].

Maier, Paul L. *In the Fullness of Time: A Historian Looks at Christmas, Easter, and the Early Church.* New York: Harper and Row, 1991.

________. "Sejanus, Pilate, and the Date of the Crucifixion." *Church History* 37 (1968): 3–13.

Maier, Paul L., ed. *Josephus: The Essential Writings.* Grand Rapids: Kregel, 1988.

Marsden, George M. *Fundamentalism and American Culture.* New York: Oxford, 1980.

Marthaler, Berard L. "The Date of Easter, Anno Domini, and Other Calendar Considerations: Chronology or Eschatology?" *Worship* 73, No. 3 (May 1999): 194–211.

Martin, Walter R. *The Kingdom of the Cults.* Minneapolis: Bethany House, 1965.

Mathematics Magazine, "Questions. Why 360 Degrees? Why Base 60?" http://www.mathematicsmagazine.com/Articles/Base60.php (accessed 23 November 2011).

Maya World Studies Center, 1996–2011. "The Maya Calendar." http://mayacalendar.com (accessed 16 November 2011).

McCarthy, Daniel P., and A. Breen. *The Ante-Nicene Christian Pasch. De Ratione Paschali: The Paschal Tract of Anatolius, Bishop of Laodicea*. Dublin: Four Courts Press, 2003.

McGowan, Andrew. "How December 25 Became Christmas." *Bible History Daily: Biblical Archaeology Review* (December 10,

2012). http://www.biblicalarchaeology.org/biblical-topics/new-testament/how-december-25-became-christmas.htm (accessed 13 December 2012).

McRoy, Anthony. "Christmas: Pagan Festival or Christian Celebration?", http://www.answering-islam.org/pagan/christmas.html (accessed 29 December 2011).

Montes, Marcos J. The Calculation of the Ecclesiastical Calendar." 2001. http://www.smart.net/~mmontes/ec-cal.html (accessed 11 November 2011).

Moonwise: New Easter Dates 1997–2205. "New Easter Dates: Proposal for astronomical calculation of Easter." http://www.moonwise.com.uk.neweaster.php (accessed 11 November 2011).

Moreland, J. P., and John Mark Reynolds, eds. *Three Views on Creation and Evolution.* Grand Rapids: Zondervan, 1999.

Moreton, Jennifer. "Doubts about the Calendar: Bede and the Eclipse of 664." *Isis* 89, No. 1 (March 1998): 50–65.

Mosshammer, Alden A. *The Easter Computus and the Origins of the Christian Era.* Oxford: University Press, 2008.

National Maritime Museum and Royal Observatory Greenwich. "The Date of Easter." http://www.nmm.ac.uk/explore/astronomy-and-time/time-facts/the-date-of-easter (accessed 11 November 2011).

Naumowicz, Josef. "La Date de Naissance du Christ d'apres Denys le Petit et les Auteurs Chrétiens Antérieurs." *Studia Patristica* XXXV (2001): 292–296.

Nicastro, Nicholas. *Circumference: Eratosthenes and the Ancient Quest to Measure the Globe.* New York: Saint Martin's Press, 2008.

O'Croinin, Daibhi, and D. Warntjes, eds. *The Easter Controversy of Late Antiquity and the Early Middle Ages: Its Manuscripts, Texts, and Tables. Proceedings of the Second International Conference on the Science of Computus in Ireland and Europe.* Turnhout: Brepols, 2012.

Ogg, George. "Hippolytus and the Introduction of the Christian Era." *Vigiliae Christianae* 16, No. 1 (March 1962): 2–18.

Oppert, Gustav. "Uber die Entstehung der Aera Dionysiana und den Ursprung der Null." *Zeitschrift für Ethnologie* (February 1900): 102–136.

Orthodox Outlet for Dogmatic Enquiries. When, Exactly, was Jesus Christ Born? http://oodegr.com/english/protestantism/imerom_genn1.htm (accessed 5 November 2011).

Peitz, Wilhelm M. *Dionysius Exiguus Studien: Neue Wege Der Philologischen und Historischen Text- und Quellenkritik. Arbeiten zur Kirchengeschichte, 33.* Berlin: Walter De Gruyter, 1960.

Ratzinger, Joseph (Pope Benedict XVI). *Jesus of Nazareth: The Infancy Narratives.* New York: Image, 2012.

"The Reason for the Season." http://mouth-shut.deviantart.com/journal.htm (accessed 30 December 2011).

Rees, Martin, ed. *Universe: The Definitive Visual Guide.* London: Dorling Kindersley, 2005.

Restall, Matthew, and Amara Solari. *2012 and the End of the World: The Western Roots of the Maya Apocalypse.* Lanham, MD: Rowman and Littlefield, 2011.

Rhoads, John H. "Josephus Misdated the Census of Quirinius." *JETS* 54, No. 1 (March 2011): 65–87.

Rich, Tracey R. "Judaism 101: Jewish Calendar." 1995–2011. http://www.jewfaq.org/calendar.html (accessed 11 November 2011).

Richards, E. G. *Mapping Time: The Calendar and Its History.* Oxford: Oxford University Press, 1998.

Rickard, Stanley Edgar, Jr. "Year of Christ's Birth." Bible Studies at the Moorings. http://themoorings.org/apologetics/chronology/Chrmas.html (accessed 6 November 2011).

Robinson, John A. T. *The Priority of John.* J. F. Coakley, ed. Eugene, OR: Wipf and Stock, 2011.

Rosetta Calendar: A Calendar Conversion Service. http://www.rosettacalendar.com (accessed 29 December 2011).

Ross, Kelley L. "The Solar Terms and the Chinese Calendar." 2010. http:// www.friesian.com/chinacal.htm (accessed 11 November 2011).

Roth, Cecil, ed. *Encyclopaedia Judaica.* 18 vols. Jerusalem: Keter/ Macmillan, 1978.

Rothwangl, Sepp. "Consideration of the Origin of the Yearly Count in the Julian and Gregorian Calendars." University Centre for Astrological Research. http://www.calendersign.com/en/to _adjustment_AD.php (accessed 11 November 2011).

Rühl, Franz. *Chronologie des Mittelalters und der Neuzeit.* (Berlin: Reuther and Reichard, 1897.

Rüpke, Jörg. *A Companion to Roman Religion* (Blackwell Companions to the Ancient World). Malden, MA: Blackwell, 2007.

Sachs, A. J., and C. B. F. Walker. "Kepler's View of the Star of Bethlehem and the Babylonian Almanac for 7/6 BC." *Iraq* 46, No. 1 (Spring 1984): 43–55.

Samuel, Alan E. *Greek and Roman Chronology, Calendars and Years in Classical Antiquity.* (*Handbuch der Altertums-wissenschaft.*) Munich: C. H. Beck'sche Verlagsbuchhandlung, 1972.

Sandeen, Ernest R. *The Roots of Fundamentalism: British and American Millenarianism, 1800–1930.* Chicago: University of Chicago Press, 1965.

Sato, Masayuki. "Comparative Ideas of Chronology." *History and Theory* 30, No. 3 (October 1991): 275–301.

"Saturnalia: The Carnaval Celebration that Became Christmas and New Year's Eve." http://www.carnaval.com/saturnalia/ (accessed 27 July 2012).

Schukowski, Manfred. *Wunderuhren: Astronomische Uhren in Kirchen der Hansezeit.* Schwerin: Thomas Helms Verlag, 2006.

Seabury, Samuel. *The Theory and Use of the Church Calendar in the Measurement of Time.* New York: Pott and Young, 1872.

Schürer, Emil. *A History of the Jewish People in the Time of Jesus Christ.* New York: Scribner, 1896.

Shepherd, Massey H. *The Paschal Liturgy and the Apocalypse.* Richmond, VA: John Knox Press, 1960.

Shepherd, Massey H., Jr. "Are Both the Synoptics and John Correct About the Date of Jesus' Death?" *JBL* 80, No. 2 (June 1961): 123–132.

The Shepherd's Page. http://www.abdicate.net (accessed 29 October 2011).

Siemon, Michael. Easter Algorithms I. http://www.cs.rutgers.edu /pub/soc.religion.christian/faq/easter-date (accessed 30 December 2011).

Smith, Mark D. "Of Jesus and Quirinius." *CBQ* 62, No. 2 (April 2000): 278–293.

Sobel, Dava. *Longitude: The Story of a Lone Genius Who Solved the Greatest Scientific Problem of His Time.* New York: Penguin, 1996.

Spencer, Bill. Blue Water Arts. "Inter Gravissimas." http://www .bluewaterarts.com/calendar/NewInterGravissimas.htm (accessed 29 December 2011).

Stark, Rodney. The Victory of Reason: How Christianity Led to Freedom, Capitalism, and Western Success. New York: Random House, 2006.

Steel, Duncan. *Marking Time: The Epic Quest to Invent the Perfect Calendar.* New York: John Wiley, 2000.

Steinmann, Andrew E. *From Abraham to Paul: A Biblical Chronology.* Saint Louis: Concordia Publishing House, 2011.

Stern, Sacha. *Calendar and Community: A History of the Jewish Calendar, Second Century BCE to Tenth Century CE.* Oxford: University Press, 2001.

Stöger, Alois. *The Gospel According to Saint Luke.* London: Burns & Oates, 1969.

Strobel, August. *Ursprung und Geschichte des Frühchristlichen Osterkalenders.* Berlin: Akademie-Verlag, 1977.

Strobel, Lee R. *The Case for a Creator: A Journalist Investigates Scientific Evidence that Points toward God.* Grand Rapids: Zondervan, 2004.

Taylor, Lily Ross. "Quirinius and the Census of Judea," *AJP* 54, No. 2 (1933): 120–133.

Thomas, Hugh. *Rivers of Gold: The Rise of the Spanish Empire, from Columbus to Magellan* New York: Random House, 2005.

Thorley, John. "When was Jesus Born?," *Greece and Rome*, Second Series, 28, No. 1 (April 1981): 81–89.

Tighe, William J. "Calculating Christmas." *Touchstone: A Journal of Mere Christianity* 16, No. 10 (December 2003): 12–15.

Touring the Mosaic City (Ravenna, Italy)—Part Two. http://www.travel-wonders.com/2009/07/touring-mosaic-city-ravenna-italy-part_10.html (accessed 30 December 2011).

U. S. Naval Observatory, Astronomical Applications Department. The 21st Century and the 3rd Millennium—When Will They Begin? http://aa.usno.navy.mil/faq/docs/millennium.php (accessed 5 November 2011).

Van Gent, R. H. Institute for History and Foundations of Science. "A Perpetual Easter and Passover Calendar." 2003. http: //www.phys.uu.nl/~vgent/easter/eastercalculator.htm (accessed 25 June 2010).

Van Goudoever, J. *Biblical Calendars.* Leiden: E. J. Brill, 1961.

Vardaman, Jerry, and Edwin M. Yamauchi, eds. *Chronos, Kairos, Christos: Nativity and Chronological Studies Presented to Jack Finegan.* Winona Lake, IN: Eisenbrauns, 1989.

Vatican City State. "Ansprache von Johannes Paul II. An die Teilnehmer der Vollversammlung der Päpstlishen Akademie der Wissenschaften. 31.October 1992." http://www.vatican.va/holy_father/john_paul_ii/speeches/1992/october/documents/hf_jp-ii_spe_19921031_accademia-scienze_ge.html (accessed 31 December 2011).

Veith, Gene Edward. "Why December 25?: The Origin of Christmas Had Nothing to Do with Paganism." *World* (December 10, 2005): 32.

Vermes, Geza. *The Nativity: History and Legend.* New York: Doubleday, 2007.

Wallis, Faith. *Bede: The Reckoning of Time.* Liverpool: Liverpool University Press, 1999.

Waltke, Bruce K., R. Laird Harris, and Gleason L. Archer. *Theological Wordbook of the Old Testament.* 2 vols. Chicago: Moody, 1980.

Ware, Steven L. "When Does the New Century Begin, Anyway?" *Lutheran Partners* 14, No. 1 (January/February 1998): 24–27.

________. "When is Easter This Year?" *Christian History and Biography*, April 20, 2000. http://www.christianitytoday.com/ch/news/2000/apr20.html (accessed 5 November 2011).

Web Exhibits. Calendars Through the Ages. 2008. http://www.webexhibits.org/calendars/calendar-ancient.html (accessed 16 February 2009).

Webb, John Charles. Light into Matter. "The Jupiter Project." http://www.templeofsolomon.org (accessed 29 December 2011).

Whisenant, Edgar C. *88 Reasons Why the Rapture Could Be in 1988.* Nashville: World Bible Society, 1988.

Whiston, William, trans. *The Works of Josephus.* Peabody, MA: Hendrickson, 1980.

White, Andrew Dickson. *A History of the Warfare of Science with Theology in Christendom.* 2 vols. New York: D. Appleton, 1896.

Williams, Frank, trans. *The Panarion of Epiphanius of Salamis.* Leiden: E. J. Brill, 1987.

Wilson, Jerry. April Fool's Day. http://www.wilstar.com/holidays/aprilfool.htm (accessed 5 November 2011).

Witcombe, Chris. "Earth Mysteries: Archaeoastronomy at Stonehenge," http://www.britannia.com/wonder/thehenge.html (accessed 30 July 2012).

Woods, Thomas E., Jr. *How the Catholic Church Built Western Civilization.* Washington: Regnery, 2005.

Wurthwein, Ernst. *The Text of the Old Testament.* Grand Rapids: Eerdmans, 1979.

Index of Scripture and Early Christian Authors

Index of Dates

Subject Index

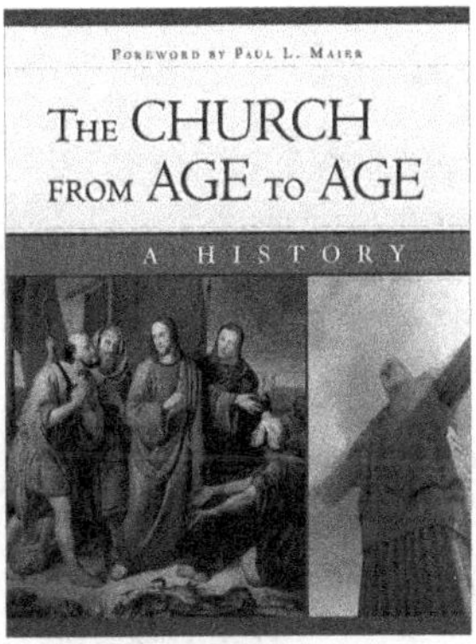

The Church from Age to Age

A History from Galilee to Global Christianity

General Editor Edward A. Engelbrecht

"An outstanding book! . . . Combines all the elements that make for a great text."
—Dr. Robert Caldwell, Southwestern Baptist Theological Seminary

The Church from Age to Age examines key historic events from the time of the apostles through today. Informative and clearly written, readers of all ages will find the answers to the who, why, and how behind the current state of Christianity the world over. Maps, readings from primary sources, and an extensive bibliography, index, and timeline make this a complete one-volume resource for the classroom and for home.

Contributors include Robert G. Clouse, Karl H. Dannenfeldt, Edward A. Engelbrecht, Marianka S. Fousek, Walter Oetting, K. Detlev Schulz, Roy A. Suelflow, and Carl A. Volz. (P) 1048 pages. Paperback; eBook.

12-4370LBR **978-0-7586-2646-2**

From Abraham to Paul: A Biblical Chronology

Andrew E. Steinmann

"Andrew Steinmann has placed biblical scholarship in his debt . . . Grounded in primary texts, Steinmann lays out here a foundation that doubtless will provide the basis for all subsequent discussions of biblical chronology, an indispensable preliminary to a proper understanding of the biblical narrative."
—Eugene H. Merrill, PhD, Dallas Theological Seminary

From Abraham to Paul presents a chronology of biblical events beginning with the life of Abraham and ending with the ministry of Paul. By discussing various arguments and providing reasoning for how we know when events happened, Steinmann presents a well-organized and coherent chronology to guide pastors, seminary students, and interested scholars through the issues in biblical chronology.

(P) 464 pages. Hardcover; eBook.

53-1185LBR **978-0-7586-2799-5**

---

www.cph.org • 1-800-325-3040

Concordia Publishing House

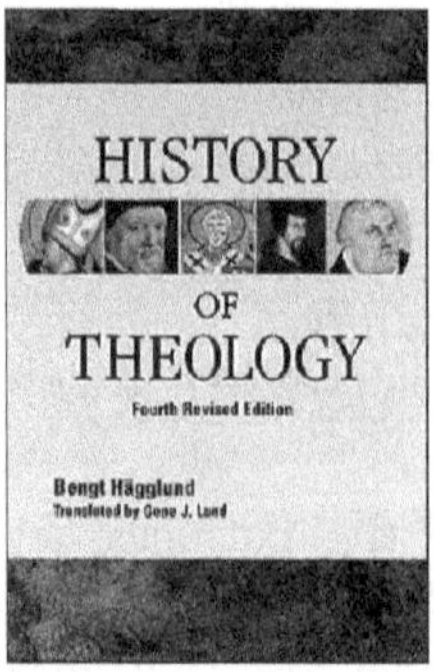

History of Theology

Fourth Revised Edition

Bengt Hägglund; translated by Gene J. Lund

History of Theology traces the movements and countermovements of theological thought through the centuries from the New Testament to the present. Part one: The Age of the Church Fathers; Part two: The Middle Ages, from Augustine to Luther; Part three: The Modern Period from the Reformation to the Present. Includes index of persons and subject index.

(P) 432 pages. Hardback. Also available as an eBook.

15-5100LBR **978-0-7586-1348-6**

The Great Works of God: The Mysteries of Christ in the Book of Genesis

Valerius Herberger; translated by Matthew Carver

"In our day so starved for a religion of the heart, Herberger opens up the contemplative study of the Word of God so that it becomes for us a word of life. He preached in an age full of change and trouble, an age that yearned for something more than the power of positive thinking. . . . The Christian Church has had a lot of great preaching—the sort of preaching that is indeed a means of grace." —Professor H. O. Old, PhD, Erskine Theological Seminary, SC

"The Lutheran pastor and hymn-writer Valerius Herberger was known as the 'Jesus Preacher' in his day, and this focus is quite evident in his very readable commentary on Genesis. Divided into 178 Meditations, each headed 'Jesus...', it combines thorough familiarity with the Hebrew language and Bible history with practical applications filled with a warm awareness of the Savior's presence in the Christian's life. . . . It will be found a rich and delightful resource for sermons and Bible classes as well as private devotions." —Professor Thomas Manteufel, PhD, Concordia Seminary, St. Louis, MO

Herberger (1562–1627), a Lutheran pastor in Fraustadt (now Wschowa), Poland, at the turn of the seventeenth century, preached through the books of the Old Testament from Genesis through Ruth, producing devout meditations on the Scriptures. "These he regarded rather like the linen cloths that wrapped the infant Jesus in the manger, and traced his Lord in every little wrinkle" (from the translator's preface). (P) Paperback; eBook.

Parts 1 & 2: Genesis 1 - 15 (424 pages). **53-1178LBR** **978-0-7586-2668-4**

Parts 3 & 4: Genesis 16 - 50 (544 pages). **53-1179LBR** **978-0-7586-2669-1**

---

www.cph.org • 1-800-325-3040

Concordia Publishing House